EVIDENCE, DECISION AND CAUSALITY

Most philosophers agree that causal knowledge is essential to decision-making: agents should choose from the available options those that probably *cause* the outcomes that they want. This book argues against this theory and in favour of Evidential or Bayesian Decision Theory, which emphasizes the symptomatic value of options over their causal role. It examines a variety of settings, including economic theory, quantum mechanics and philosophical thought-experiments, where causal knowledge seems to make a practical difference. The arguments make novel use of machinery from other areas of philosophical inquiry, including first-person epistemology and the free-will debate. The book also illustrates the applicability of decision theory itself to questions about the direction of time and the special epistemic status of agents.

ARIF AHMED is Senior Lecturer in Philosophy at the University of Cambridge.

EVIDENCE, DECISION
AND CAUSALITY

ARIF AHMED

University of Cambridge

CAMBRIDGE
UNIVERSITY PRESS

University Printing House, Cambridge CB2 8BS, United Kingdom

Cambridge University Press is part of the University of Cambridge.

It furthers the University's mission by disseminating knowledge in the pursuit of education, learning and research at the highest international levels of excellence.

www.cambridge.org
Information on this title: www.cambridge.org/9781316641545

First published 2014
First paperback edition 2016

A catalogue record for this publication is available from the British Library

Library of Congress Cataloguing in Publication data
Ahmed, Arif.
Evidence, decision, and causality / Arif Ahmed, University of Cambridge.
pages   cm
Includes bibliographical references and index.
ISBN 978-1-107-02089-4 (hardback : alk. paper)
1. Decision making.   2. Evidence.   3. Causation.   I. Title.
BD184.A36   2014
122 – dc23   2014020933

ISBN 978-1-107-02089-4 Hardback
ISBN 978-1-316-64154-5 Paperback

Contents

Preface

Causality is a pointless superstition. These days it would take more than one book to persuade anyone of that. This book focuses on the 'pointless' bit, not the 'superstition' bit. I take for granted that there are causal relations and ask what doing so is good for. More narrowly still, I ask whether causal belief plays a special role in decision. My argument that it does not consists largely of schemes for extracting money from the people who think it does. I conduct the argument using the framework of decision theory. Decision theory is convenient for this purpose because it (a) quantifies causal beliefs and (b) isolates their role in practical deliberation.

Because of this aim the book couldn't serve as a stand-alone introduction to decision theory. Anyone who took it that way would find it distorted and lacunary. Distorted because of the focus on the dominance principle at the expense of almost everything else that Savage's axioms entail; lacunary because of the complete absence from the story of any approach outside the Ramsey–Savage expected-utility paradigm.

Anyway, many excellent introductions to the subject are already available at various levels of mathematical sophistication. For instance, Peterson 2009 is written at a mathematically elementary level, Gilboa 2009 and Kreps 1988 are more difficult and Fishburn 1970 is mathematically fairly advanced. *This* book presupposes no mathematical knowledge beyond completely elementary set theory and probability.

But it probably *could* function as a philosophical companion to any such introduction. Its central topic is perhaps the main debate in the philosophical foundations of decision theory. That subject has a bearing on more traditional preoccupations of metaphysics, including causality itself, the asymmetry of time and the nature of self-knowledge. Pursuing it also forces us to touch upon live issues in psychology, economics, the theory of voting and the foundations of quantum mechanics. So I hope in these chapters to illustrate why the philosophy of decision theory is important, or at least interesting, even if you are not a philosopher of decision theory.

I owe the reader an apology for the number and the length of the footnotes. They largely involve (i) references, (ii) numerical calculations and (iii) objections and clarifications that are unlikely to interest most readers but likely to interest some. In cases (ii) and (iii), I felt that including these items in the main text would break up the main thread of the argument, which you should be able to follow without reading any of them.

Whilst writing this book I have had the benefit of conversations with, and/or written comments from, the following people: Helen Beebee, Sharon Berry, Simon Blackburn, Rachael Briggs, Lucy Campbell, Adam Caulton, John Collins, Tom Dougherty, Adam Elga, Luke Fenton-Glynn, Alison Fernandes, Alexander Greenberg, Alan Hájek, Caspar Hare, Jane Heal, Hykel Hosni, Jennan Ismael, Leon Leontyev, Isaac Levi, Hanti Lin, Penelope Mackie, John Maier, Adam Morton, Daniel Nolan, Harold Noonan, Huw Price, Paolo Santorio, Wolfgang Schwarz, Shyane Siriwardena, Julia Staffel and Paul Weirich. I thank them all. I also thank two referees from the Press for their extremely helpful written comments, Dr John Gaunt for meticulous copyediting, and Alexander Greenberg and Shyane Siriwardena for assistance with the index. And I thank my editors at CUP, Sarah Green and Hilary Gaskin, for their efficiency, patience and goodwill.

I am also grateful to audiences at the Aristotelian Society, the University of Auckland, the Australian National University, Birmingham University, Bristol University, the University of Cambridge, Columbia University, the Institute of Philosophy, Monash University, the University of Nottingham, Princeton University, the Scuola Normale Superiore di Pisa, the University of Sheffield, the University of Sussex and the University of Sydney, where I delivered talks on material that has ended up in the book.

I wrote some of this book whilst holding a Leverhulme Research Fellowship at the Faculty of Philosophy, University of Cambridge, and then at the Sydney Centre for the Foundations of Science, University of Sydney. I wrote some of it whilst holding a Distinguished Visiting Professorship at the Scuola Normale Superiore di Pisa. And I wrote some of it whilst holding a Visiting Fellowship at the Research School of Social Sciences in the Australian National University, Canberra. I am grateful to all of these institutions for their hospitality, and also to the Leverhulme Trust. I also thank the Faculty of Philosophy at the University of Cambridge, and Girton College, Cambridge, for granting me leave from teaching and administrative duties in order to finish the book.

Material from section 3.1 has appeared in my paper 'Push the button', *Philosophy of Science* 79 (July 2012): 386–95, and is reproduced by permission of the University of Chicago Press. Copyright 2012 by the Philosophy of Science Association. All rights reserved.

Material from section 5.1 has appeared in my paper 'Causal Decision Theory and the fixity of the past', published online in the *British Journal for the Philosophy of Science* on 16 September 2013 by Oxford University Press on behalf of the British Society for the Philosophy of Science.

Material from section 5.2 has appeared in my 'Causal Decision Theory: a counterexample', in *Philosophical Review*, 122, 2: 289–306. Copyright 2013, Cornell University. Reprinted by permission of the present publisher, Duke University Press, www.dukepress.edu.

Material from s. 7.2 is here reproduced from my paper 'Infallibility in the Newcomb Problem', published online in *Erkenntnis* on 11 April 2014, copyright Springer Science+Business Media Dordrecht 2014, and appears here with kind permission from Springer Science+Business Media.

Material from section 7.3.3 has appeared in A. Ahmed and H. Price, 'Arntzenius on "Why Ain'cha rich?"', *Erkenntnis* 77: 15–30, copyright 2012 Springer Science+Business Media B.V., and reappears here with kind permission from Springer Science+Business Media B.V.

Finally, and most of all, I am grateful to my family (Frisbee, Isla, Iona and Skye) for putting up with me during the writing of this book. My father, Dr G. M. Ahmed, died in early 2013 after a period of illness that he bore with characteristic stoicism and good humour. I should like to dedicate the book to my mother, Mrs S. Ahmed, and to the memory of my father.

Introduction

0.1 Causalism and evidentialism

Causalism is the doctrine that rational choice must take account of causal information. Specifically it must attend to whether and how an agent's available acts are *causally relevant* to the outcomes that he desires or dreads. Nothing short of causal information will do the trick, and agents who ignore it will make the wrong decision in a variety of identifiable and realistic cases. This book argues against causalism.

Evidentialism, which it prefers, is the contrary view that only the *diagnostic* bearing of acts is of practical concern. It only matters to what extent this or that act is *evidence* of this or that outcome, regardless of whether the act causes the outcome or is merely symptomatic of it. Many cases where evidentialism and causalism make different practical recommendations are less realistic than causalism supposes. But anyway evidentialism is practically superior: we do better to follow the evidentialist recommendation than to follow the causalist one, *whenever* they diverge.

Two philosophical positions encourage evidentialism. The first is Humean scepticism about causality.[1] I strike the match and then it lights. However closely you examine that sequence you won't see the striking *making* the match light. However often you examine similar sequences you will only see the recurrence of that pattern. However widely you examine surrounding sequences you will only see recurrences of other patterns. All of this only shows that in such-and-such circumstances the striking of a match is a good *sign* of its lighting. So as far as we can see, distinct events are, as Hume said, entirely loose and separate. So causality does not impinge upon human experience in any other way than through the *evidential* relations between events (the patterns) that either constitute causality or are sustained by it. This is not to say that some deeper metaphysical relation of causality

---

[1] Hume 1949 [1738]: I.iii.

1

doesn't exist, but only that for practical purposes we needn't care what it is or whether it does.

The second motivation is Russellian eliminativism about causality.[2] Physics alone tells us what really causes what. There are no causes in physics. So there are no causes. To act as if there were is to act upon illusions. To act upon illusions is as counterproductive in the practical sphere as believing them in the theoretical sphere.

Causalism can seem attractive from the perspective of an anti-Humean tradition pre-dating Hume himself. According to it, Hume misses out half the story. In fact we are not just *observers*, but also *agents*. Therefore we must acknowledge aspects of reality that a mere spectator might have no reason to discern. In its most romantic and excitable version (Schopenhauer's), *only* the willing of an agent can penetrate the veil that conceals *all* objective truth from mere perception. But you needn't go that far to appreciate the insight, if it is one, that rational agents have reason as such to distinguish tighter *causal* relations between events from amongst the merely evidential relations of co-occurrence and statistical correlation that are equally available to Humeans and to the paralytic.[3] Still, many modern adherents of causalism make no reference to that tradition.

It is possible to question all of these standpoints. In connection with Humean scepticism, one might object that people can in fact *see* e.g. that one thing is pushing another.[4] In connection with Russellian eliminativism, one might deny both (a) that only physics can settle what causes what[5] and (b) that physics makes no room for causality.[6] And in connection with the 'anti-Humean' tradition, one might object that agency, being itself a causal notion, cannot reveal causality to anyone who would otherwise be innocent of it.[7]

But my aim in mentioning these metaphysical positions (for which plenty more might be said) wasn't to argue for evidentialism or causalism, but only to indicate what broader issues give additional interest to the dispute between them. My reasons for preferring evidentialism don't depend on these sweeping metaphysical arguments, but rather on attention to individual cases where the choice between these doctrines makes a difference.

---

[2] Russell 1913.
[3] The tradition includes Berkeley (1980 [1710]: sections 25–9); Reid (2001 [1792]); Schopenhauer (1995 [1819]: Book II); Collingwood (1940: 291, 322); Von Wright (1973: 108–13); and Menzies and Price (1993: 191–2).
[4] Michotte 1963.   [5] Cartwright 1979.   [6] Hitchcock 2007.
[7] Ahmed 2007 discusses some forms of this objection.

So as to convey what that involves, the rest of this Introduction sets out in informal terms: (0.2) the kind of evidential relation and (0.3) the kind of causal relations that will matter here; (0.4) the formal framework in which I propose to conduct the debate; (0.5) an explication of the opposing positions in terms of it; and (0.6) the kind of case over which they disagree. Section 0.7 outlines the relevance of these matters to philosophers and others for whom they may not hold intrinsic interest.

0.2 Evidence

Evidence is here an (a) subjective and (b) non-causal relation that I'll take (c) to hold between propositions. (a) Whether one thing is evidence for another may vary across persons. (i) You might think that Pete's calling in this hand of poker is evidence that he'll lose; I think it is evidence that he'll win. (ii) I might think that the report of a miracle is evidence of a divine plan whereas you think it is irrelevant.

Such disagreement needn't imply irrationality on either side. What a rational person takes as evidence itself depends upon (i) particular and (ii) theoretical background knowledge or belief, and either may vary interpersonally. (i) Suppose you know that Pete has a worse hand than his opponent's; but I know that Pete has sneaked a look at his opponent's hand. (ii) Suppose my religion says that if God exists then He regularly intervenes in human affairs; but according to yours, the regular course of nature is the only possible expression of His will.

So evidence is subjective. And I won't assume any objective constraints on what can count as evidence for what, beyond the dictates of logic and probability (see sections 1.1, 1.4 and 2.3). On the other hand, there is a limit to what we can learn by considering agents with utterly outlandish beliefs on this point, as the literature on this subject sometimes does (see section 7.1). For this reason I think that the cases worth most attention involve agents whose beliefs you could at least imagine holding. Chapters 4–6 investigate these.

(b) Evidence does not only relate causes to their effects. Of course causes *can* be evidence of effects: my turning the key in the ignition is evidence that the car is going to start. But this is not the only case. (i) This bloodstained knife is evidence that the butler did it. But its being bloodstained is an effect and not a cause of the crime. (ii) The bloodstained knife is also evidence that the butler will confess under questioning. But it may be neither cause nor effect of this. Instead, the deed itself might be a common cause of both. (iii) Certain theories of historical development or of human nature could have led an educated observer of the French Revolution quite reasonably to expect it to develop along the same lines as the English one that occurred a century before. So the past course of the English Revolution is, for him, evidence of the future course of the French Revolution. But the observer might deny *both* that any causal connection relates these sequences *and* that some particular past state or event caused both of them.[8]

(c) In these examples, and also in ordinary speech, many kinds of thing count as subjects or objects of 'evidence', e.g. events (Pete's calling), objects (a bloodstained knife), states of affairs (the existence of God) or historical sequences (the French Revolution). There may be nothing wrong with that. But here I shall for convenience say that it is *propositions* that evidentially support other *propositions*. Instead of saying that the bloodstained knife was evidence of the butler's guilt, I'll say *that the knife was bloodstained* is evidence *that the butler did it*. Similarly I will say that *whether the knife was bloodstained* is evidentially relevant to *whether the butler did it*, meaning by this that the proposition that the knife was bloodstained has some evidential bearing, positive or negative, on the proposition that the butler did it.

0.3   Causality

The causal relations that will matter are *causal dependence* and *independence* between possible individual events, facts or states of affairs – or rather, again, between particular propositions. For instance, whether this bottle smashes is causally dependent on whether it is dropped but causally independent of whether that match is struck. Causal relevance is the converse of causal dependence. Whether the match is struck is causally relevant to whether it lights but causally irrelevant to whether the bottle smashes.

Except in Chapter 6, it will be clear enough in most cases what causally depends on what. But it's worth stating that E can causally depend on C even though it wouldn't be natural to say that C caused E, or that C was 'the' cause of E. For instance, the lighting of this match was causally dependent on the presence of oxygen in the atmosphere. But it sounds odd to say that the presence of the oxygen caused the match to light. That is not because

---

[8] Conversely, something can be a cause without also being a sign. My striking of this particular match might ultimately cause the room to fill with cigarette smoke. But it isn't evidence that the room will fill with smoke, at least not to anyone who knows that I also have a lighter or can borrow a light from any of a dozen other people, to which person it was already practically certain that the room will fill with smoke whether or not I strike that particular match. But *this* misalignment between the evidential and the causal has little relevance here: what matters is the difference between signs that are causes and signs that are not causes, not the difference between causes that are signs and causes that are not signs.

the oxygen bears *no* causal relation to the lighting but because we select for largely subjective reasons a single event, state or fact, from amongst the many on which the lighting was causally dependent, that we then call '*the*' cause. Causalism doesn't care about the things of which my act may be the cause in this narrower sense, but only about what may causally depend on it in the 'broad and non-discriminatory' sense.[9]

I will take for granted that there *is* such a thing as causal dependence, i.e. that our use in deliberation of causal vocabulary picks out a single metaphysical relation, that that relation is causal dependence, and that sometimes some things really *are* causally dependent on other things. Not because I have any great confidence in that thesis, being myself attracted to the view that not only rational decision but also physical science can manage quite well without it, but because causalism itself presupposes it.

Nor will I raise any concerns, Humean or otherwise, about the episte-mology of causal dependence. I take for granted that the normal procedures that we take to establish causal dependence – e.g. controlled trials – do give us reason to think that whether this particular event occurs is causally dependent on whether that other one does. This is not because such doubts are straightforwardly answerable but because they are irrelevant to the assessment of causalism. The question is not whether causation exists or how we can know about it; the question is why we should care.

0.4   Decision theory

Locke wrote:

> What we once know, we are certain is so: and we may be secure, that there are no latent proofs undiscovered, which may overturn our knowledge, or bring it in doubt. But, in matters of probability, it is not in every case we can be sure that we have all the particulars before us, that any way concern the question; and that there is no evidence behind, and yet unseen, which may cast the probability on the other side, and outweigh all, that at present seems to preponderate with us . . . And yet we are forced to determine ourselves on the one side or other. The conduct of our lives, and the management of our great concerns, will not bear delay: for those depend, for the most part, on the determination of our judgement in points, wherein we are not capable of certain and demonstrative knowledge, and wherein it is necessary for us to embrace the one side, or the other.[10]

---

[9] E.g. Lewis (1973a: 162) distinguishes this sense and implicates it in his own version of causalism (1981a: 329–35).

[10] Locke 1975 [1689]: IV.xvi.3.

The second half of this contrast ('And yet . . . ') is wrong. The conduct of our lives does *not* force us to embrace one side or the other. We often act, and sometimes act rationally, without any definite opinion on points that matter to the outcome.

This coin lands heads two tosses out of three. I must make a bet that pays my stake if the coin lands heads this time. It is certainly rational for me to put *some* but not *all* of my present wealth on heads. But when I put up the stake I neither have nor feign certainty either that the coin will land heads (otherwise I'd stake *all* of my present wealth) or that it won't (otherwise I'd stake *none* of it). By staking some intermediate amount I am acting rationally without embracing one side or the other.

Normative decision theory tries to say in quite general and abstract terms just *how* to act under such uncertainty. This book conducts the dispute between evidentialism and causalism in the terms of normative decision theory, which from now on I'll just call 'decision theory'. I shall largely set aside *descriptive* decision theory, which tries to make general claims about how agents' *actual* behaviour depends on their values and beliefs.

Decision theory works roughly like this. Suppose an agent has several options in some situation. The situation has many possible outcomes that the agent desires or dreads in varying degrees. For each option, if she takes it, some outcomes are more likely than others. At any rate she *takes* some outcomes to be more likely than others if she takes the option. Suppose we have some idea of *how* good each outcome is for her. And suppose we have some idea of *how* likely she considers each outcome, given each option. Decision theory takes all this as input – the options, the possible outcomes, and how good and how likely she considers them. Its output is a *recommendation* of some option or options. *Contra* Locke, you *can* act rationally in the absence of real or feigned certainty, and decision theory tells you how.

Thus suppose that the agent can bet any positive fraction of her current fortune on the next toss of a coin whose chance of landing heads is in her view definitely two-thirds. If the coin lands heads then she is better off by the amount of her stake. If the coin lands tails then she is worse off by the same amount. What fraction of her fortune should she bet?

The point of decision theory is to answer questions like that. In this case the agent has many options (one for every amount that she might stake). And there are many possible outcomes: for each dollar amount k that she bets out of a dollar fortune F, there is the good outcome that she ends up with $\$(F + k)$ (if the coin lands heads) and the bad outcome that she ends up with $\$(F - k)$ (if the coin lands tails). We know that her confidence in heads is two-thirds. If we also know the rate at which she values each extra dollar, then decision theory should tell her what to do. And on a simple if not entirely plausible assumption about the rate at which she values money, it turns out that most sensible decision theories tell her to stake one-third of her fortune.[11]

Not every possible decision theory *is* sensible. Consider the theory that tells you always to do what is most costly: that is a stupid theory. So is the one that tells you always to do what is most risky. So is the one that tells you always to do what is *least* risky. In fact infinitely many decision theories make palpably absurd recommendations in all sorts of cases. I'll ignore all of them.

More seriously, I'll largely ignore *objective* decision theories. An objective decision theory specifies the best option independently of the *agent's* uncertainty about the state of the world. Suppose e.g. that at each time each possible future event has an *objective chance*. The objective chance of an event at a time is some quantity that is independent of anyone's confidence that it will occur: for instance, the chance that this radium atom here will decay in the next minute. There are objective decision theories whose outputs depend on chances themselves, and not on the agent's beliefs about chances. One such theory advocates doing *whatever now has the best chance of bringing about the best outcome*, quite independently of what the agent thinks about this. Section 3.3 below states my reasons for setting aside objective theories.

I'll focus instead on theories into which the *facts* of the agent's situation enter *only* in so far as her beliefs reflect them. Such decision theories, unlike objective ones, generally give usable advice to agents that use them. They are *subjective* decision theories. And I'll focus for the most part on the two subjective theories that philosophers have focused upon for the last forty years. These are *Evidential* (sometimes also called *Bayesian*) *Decision Theory* (EDT) and *Causal Decision Theory* (CDT). Neither one is absurd. But they can't both be true because they sometimes disagree.

0.5 Evidential Decision Theory and Causal Decision Theory

EDT and CDT are both quantitative theories. They depend upon some numerical measure of the value of each outcome for the agent and also of the uncertainty that she attaches to relevant hypotheses about the state of the world. So I can't explain how they work until I've explained how

---

[11] The assumption is logarithmic utility.

to quantify both value and uncertainty. Chapters 1 and 2 do that in more detail. Here I explain only the general idea behind each theory.

EDT identifies the value of an option with its *news* value. It recommends what is in the agent's view the most *auspicious* option, the one that good fortune most probably *accompanies*. CDT identifies the value of an option with the value of its believed *effects* (including itself). It recommends the option that the agent considers most *efficacious*, that most probably *produces* good fortune.

It is important to distinguish EDT from 'magical thinking': the false belief that one can *causally* influence the outcome of some process by symbolic gestures or other indirect means.[12] People who suffer from illnesses can't, for the most part, cure themselves by *acting as if* they have recovered. Voting for your candidate doesn't *cause* other like-minded people also to vote for him. EDT can certainly acknowledge that in these cases there is absolutely no *causal* connection between the act and the desired state. But then neither, in these cases, is an agent likely to see much *evidential* connection between them. EDT only insists that when the agent genuinely *does* take some act to be good news, he has reason to perform it.

Even this vague exposition of their differences reveals EDT and CDT as versions of evidentialism and causalism respectively. The news value of an act is a function of its evidential bearing upon outcomes of interest, irrespective of whatever causal relations lie beneath. So EDT cashes out the central commitment of evidentialism: that only its diagnostic or evidential import need be relevant to the assessment of an act. Similarly, CDT crystallizes the central idea of causalism, that the practical value of an act is sensitive to its causal bearing upon outcomes of interest. There *are* some causalists who think that CDT does not correctly spell out the nature of the sensitivity, and I will discuss three arguments to this effect in Chapter 3. But CDT is the most popular version of causalism.[13] In what follows I shall for the most part identify the issue between evidentialism and causalism with the issue between EDT and CDT.

Putting them in the way that I have, both decision theories sound reasonable. At least neither is as absurd as the theory that always recommends the most costly option, or the one that always recommends the least risky

---

[12] Shafir and Tversky 1992: 463.
[13] Causalist defenders of CDT include Nozick (1969: 222 ff.); Gibbard and Harper (1978: 355–7); Lewis (1981b: 308–12); Joyce (1999: Chapter 5); Pearl (2000: 108–10); Sloman (2005: Chapter 7); and many others. Causalists who reject CDT, or at least don't plainly endorse it, include Cartwright (1979; see Lewis 1981b: 325 n. 15); Egan (2007; see section 3.1 below); and Mellor (1983, 2005; see Section 3.3 below).

option. The way to settle the issue between them is to look at situations where they conflict. There a contemplated act typically has an *evidential* bearing on an outcome that it *does nothing to bring about*. We have already seen (at section 0.2) possible situations where *A* is evidentially but not causally relevant to *B*. What we seek now is a case of this sort in which *A* is an option and *B* an outcome to whose occurrence the agent is not indifferent.

It is a surprising fact that such cases are fairly difficult to find. In fact the most widely discussed case, which I cover at section 2.5 and also in Chapter 7, is explicitly science-fictional. Variants on this example involve stories about God, Satan, angels, fantastically powerful computers and unusually able psychologists.

Whilst its being thus hypothetical makes it attractively simple, it also raises a concern about relevance. If *in practice* EDT and CDT never disagree, then the disagreement between them might seem relatively trivial. If so, we have reason to prefer evidentialism, which says that an agent *needn't* care about the causal relevance of his options once their evidential bearing is firmly in view. I discuss this point briefly at the start of Chapter 4 and in more detail at section 7.1. In any case, and as Chapters 4–6 argue, disagreement between the theories is feasible and, in at least some cases, reasonably realistic. But given the present purposes of introduction and illustration, it is worth citing a clear and straightforward example of disagreement between EDT and CDT, even if it is as fanciful as I expect most readers to find the following.[14]

0.6 Predestination

Predestination was historically the most important feature of Calvinism.[15] Here it is, in the Westminster Confession of 1647:

> By the decree of God . . . some men and angels are predestinated unto everlasting life, and others foreordained to everlasting death. Those of mankind that are predestinated unto life, God before the foundation of the world was laid, according to his eternal and immutable purpose, and the secret counsel and good pleasure of his will, hath chosen . . . out of his free grace and love, without any foresight of faith and good works, or perseverance in either of them, or any other thing in the creature as conditions or causes moving Him thereunto. The rest of mankind God was pleased . . . to pass by. All those

---

[14] Resnik (1987: 112) calls it a real-life case, presumably because many real people genuinely believed themselves to be in it.

[15] Weber 1992 [1920]: 57.

Table 0.1 *Calvinist problem*

	Salvation	Damnation
Virtue	2	0
Sin	3	1

whom God hath predestined unto life, and those only, He is pleased . . . to call by His word and spirit . . . renewing their wills, and by His almighty power determining them to that which is good. As for those wicked and ungodly men . . . He not only with-holdeth His grace . . . but sometimes also withdraweth the gifts which they had and . . . gives them over to their own lusts, the temptations of the world and the power of Satan.

Nothing that anyone does can *bring about* his salvation. For instance, nothing that I do now can somehow *cause* God to have foreseen that I would do it and to decide my salvation on that basis: for God has already settled that '*without any foresight* . . . or any other thing in the creature as *conditions or causes* moving Him thereunto'. But although the doing of good works is not a *cause* of salvation, it is a *sign* of salvation, because if God has chosen to save me then he has already *determined* me to do good works. (Note that only incompatibilists about free will would think that this derogates from my freedom to choose in this matter.) Similarly, a sure sign of damnation, though again not a cause of it, is indulgence in one's own lusts etc. In Weber's summary: 'however useless good works might be as a means of attaining salvation . . . nevertheless, they are indispensable as a sign of salvation'.[16]

Let some Calvinist agent believe this. Imagine him facing what he considers the decisive temptation. If he yields then his life (on Earth) will be pleasurable. But yielding is a sure sign that he was damned for all eternity and so will suffer everlasting death after this life. If he declines the temptation then his life on Earth will certainly be dull. But declining is a sure sign of everlasting superlunary life.

Table 0.1 summarizes the possible outcomes and their relative value to the agent. The entries in the top row represent God's possible decrees: that the agent is saved or damned. The entries in the left-hand column represent the agent's options: to decline the temptation and to yield to it. The body of the table has four cells. Each corresponds to one of four possible outcomes.

[16] Weber 1992 [1920]: 69.

For instance, the top left-hand cell corresponds to the outcome that the agent is virtuous and is saved. And the bottom right-hand cell corresponds to the outcome that the agent is sinful and is damned. The number inside each cell represents the agent's evaluation of the outcome that it represents. At this point you should interpret the numbers as follows: for any two outcomes, the agent *prefers* the one with the *bigger* number.

So what Table 0.1 tells us about the agent's preferences over the four outcomes is just what you would expect. He prefers each outcome involving salvation to both outcomes involving damnation. And he prefers sin under any dispensation to virtue under that same dispensation. The best outcome (which scores 3) is the bottom left-hand box: the agent gets to sin *and* to be saved. The next-best outcome is the top left-hand box: salvation more than compensates for his forgoing of worldly pleasures. The third-best outcome is the bottom right-hand box. Here, sin is insufficient compensation for damnation: he would rather have the salvation and forgo the sin. Finally, the worst outcome is associated with the top right-hand box: the agent is virtuous but still damned. And the problem for the Calvinist agent is: given all these preferences, should he take the path of virtue or the path of sin? (I'm not yet concerned to justify an answer, but only to illustrate the EDT/CDT conflict.)

Virtue is a sign of salvation. Sin is a sign of damnation. So the agent's ultimate fate is *evidentially dependent* on his present choice. The agent prefers virtue and salvation (the top left-hand cell) to sin and damnation (the bottom right-hand cell). So, according to EDT, the agent should choose virtue.

But CDT advises sin. God's decree for the agent took place many years ago, or perhaps is outside time altogether. Either way, no good works can now *make* the agent's next life go any better. And sin cannot make it go any *worse*. But sin would make *this* life better for the agent than virtue would, whatever God has decreed. So the agent should be sinful.

This is a clear if unrealistic case where EDT and CDT disagree. It is a version of *Newcomb's problem*, which has attracted plenty of attention from philosophers since about 1970 and which in various realizations gets plenty here too, at sections 2.5, 4.2–6 and 5.1, and throughout Chapter 7. The strange thing about Newcomb's problem is that there is widespread agreement that it is obvious what the agent should do, but not so much over what that is.

Many people think it obvious that the agent should sin, since whether he sins has no effect on whether he is saved. Obviously, if virtue were to *cause* his salvation, e.g. by causing God to have foreseen his virtue, then the pay-offs in Table 0.1 would make virtue rational. But it doesn't, so they don't. He may as well have fun before meeting a future that would have been his whatever he had done.

Others find it completely opaque just what *about* the causal relation could make this difference. After all, the situation is practically indistinguishable from one in which virtue really does cause salvation. Everyone who is virtuous is saved, including the agent. Everyone who sins is damned, including the agent. That connection falls short of *being* causal, but how could genuine causation add anything of practical relevance?

It couldn't – not here and not anywhere else. At least not according to EDT, not according to the evidentialism that it represents, and not according to Ramsey when he wrote that 'all the practical man wants to know is that all people who take arsenic die, not that this is a causal implication, for a universal of fact is *within its scope* just as good a guide to conduct as a universal of law'.[17] Evidential Decision Theory is essentially a vast generalization of Ramsey's point. This book does not defend that theory against all objections but only its superiority to CDT, since it is on that point that the dispute between evidentialism and causalism seems to me, as it has seemed to many, to turn.

Chapter 2 presents EDT and CDT in slightly greater detail as rival developments of Savage's original decision theory, as outlined in Chapter 1. My reasons for proceeding this way are, first, that Savage's is still the most widely known and applied decision theory and so represents a convenient point of entry to this philosophical debate for decision theorists who are not familiar with it; and second, that Savage's theory represents an agreed background against which the formal difference between EDT and CDT stands out more clearly.

The main argument of the book is, then, for these two premises:

(1)   If causalism is true then the causal bearing of an act is relevant to its assessment in the way that CDT claims.

(2)   The causal bearing of an act is not relevant to its assessment in the way that CDT claims.

It follows from (1) and (2) that causalism is false. Chapter 3 argues for (1) against those who assert that the causal bearing of one's acts *is* relevant to their assessment, only not in the way that CDT describes. Chapters 4–7 argue for the second premise by examining cases where EDT and CDT

---

[17]   Ramsey 1990 [1928]: 144. His later paper on this subject (1990 [1929]) takes a more nuanced though still evidentialist line, some elements of which I outline and criticize in Chapter 8.

seem to disagree. I argue that in every such situation, *either* they really give the same advice *or* EDT gives better advice. Either way, the examples all support premise (2) and also the superiority of EDT over CDT.

Chapter 8 considers an alternative 'dualistic' version of evidentialism that seeks to defuse the debate. That view assigns a traditionally anti-Humean significance to a contrast between the *agent's* perspective and the *observer's* perspective. And when EDT accommodates this point, it turns out not to disagree with CDT anywhere. I reject both this conclusion and also its philosophical starting point. There *is* a genuine dispute between EDT and CDT. And for the purpose of evaluating his options, an agent need *not* see any deeper significance in the distinction between those bodily movements and other events that occur because he wants them to, and those on the other hand that take place willy-nilly.

0.7 Why it matters

The relevance of the dispute between Evidential Decision Theory and Causal Decision Theory outruns its bearing on causalism. It should interest (i) non-philosophers who care about decision theory, (ii) non-decision theorists who care about philosophy, and (iii) everyone else.

(i) First, decision theory standardly plays a more *descriptive* than normative role in those sciences, like economics and political science, that make widespread use of it; whereas this book is largely concerned with the foundations of the normative theory. Nevertheless, normative questions should matter to those sciences in so far as they are concerned to map human conformity to, or deviation from, ideals of practical *rationality*. For instance: when people vote in large elections, are they acting rationally or irrationally? The answer depends in part on what counts as acting rationally. So if EDT and CDT disagree about that then they disagree about which areas of human conduct are deviations from that ideal.[18]

Second, the argument of this book covers various misunderstandings of what EDT in fact recommends. For instance, the theory does *not* recommend that you shouldn't visit the doctor for fear that this will confirm that you are ill, or that you shouldn't read the newspaper for fear that you will get bad news; more generally, it does not always advise against actions that are statistically associated with bad outcomes (see section 4.1). Chapter 4 attempts to identify where and why it *does* proscribe options that bring bad news without causing bad outcomes. And

---

[18] For further discussion of the voting case see section 4.6.3 below.

whilst that point is here put in service of a normative doctrine, it also has implications for the descriptive version of EDT, which predicts the behaviour that its normative counterpart recommends. Once we understand the true consequences of EDT we are better placed to assess its descriptive bearing on the choices that people do in fact make; hence also to appreciate its potential relevance to predictive or explanatory projects in social science.[19]

Decision theory as applied in the social sciences does not normally recognize the EDT/CDT distinction. What gets much more usually applied is Savage's original theory, which is a common ancestor of EDT and CDT in a sense that section 2.8 spells out. But this is not because the difference between EDT and CDT is irrelevant to those sciences but because the general application of the Savage theory *already presupposes* a stand on the EDT/CDT dispute. For instance (and putting it in terms that Chapter 1 explains): the application of the Savage theory to the Calvinist's problem demands a decision on whether the appropriate partition of the state space is (a) into cases where you are damned and cases where you are saved, or (b) into cases where you *now act* as God decreed and cases where you now act against God's decree concerning your action. Thinking of the problem in terms of the (a)-possibilities is *already* to settle the issue in favour of CDT; thinking of it in (b)-like terms is to settle it in favour of EDT.[20]

(ii) Turning to its strictly philosophical interest: first, at least some of the cases I cover demand a stand on matters of wider philosophical interest. For instance, what you think evidentialism prescribes in a variety of cases may depend on your views on self-knowledge, in particular on your sympathy for a broadly Cartesian auto-epistemology: see sections 4.3.2 and 4.5 for further discussion.

More generally, subjective decision theory should take account of the sense, if any, in which a deliberating agent's beliefs, desires and intentions are transparent to her. So the foundations of decision theory constitute a novel field in which to apply work in this live area of philosophy of mind. Here those questions arise most prominently at sections 3.2 and 8.2.

Second, some of the cases that I discuss are themselves relevant to metaphysics. At least, they are for pragmatists: in particular those who want to know what ways of thinking about time, or about causality, best serve our practical interests. Time: section 5.4 argues that a rational decision-maker thinks of the past as open to deliberation much as everyone thinks of

---

[19] For an extensive working out of the applications of EDT to questions in political science, see Grafstein 1999.

[20] For a more realistic example that makes the same point, see Joyce 2002: 71–5.

the future as being thus open. For only this view of openness is consistent with a rational approach to the decision problems that sections 5.1 and 5.2 discuss. Causality: section 6.5.1 argues that in some cases what the *causal* theory recommends turns on what causality is, making the Chapter 6 cases of interest even to those whom they don't convince to abandon CDT.

(iii) Normative decision theory distinguishes rational from irrational behaviour against a fixed background of beliefs and desires. Although the argument in this book presupposes some idealization as outlined at section 1.1, still in many cases EDT's advice holds good even for agents that do not meet these standards. For instance, situations of the 'Prisoners' dilemma' type arise frequently, and whilst EDT largely agrees that it is more rational to defect than to co-operate in these cases, section 4.6.3 identifies marginal cases where co-operation may be rational. Marginal they may be, but it is easy enough to see how they might arise in real life. And when they do, the argument here gives practical advice about what to do in them: for instance, and continuing the example in (i) above, when it would be rational to vote in a large election.

So the subject is not only of interest to philosophers who study decision theory. But whatever your reason for approaching this subject, the proper starting point is the first systematic and still the most influential treatment of decision under subjective uncertainty, viz Savage's theory.

CHAPTER I

Savage

1.1 Simplifications and idealizations

Neither decision theory nor its philosophy deals directly with real life, but rather (and like any science) with a model that (a) simplifies and (b) idealizes it. 'Simplify' means 'ignore irrelevant facts about real people'. 'Idealize' means 'imagine away some of their actual limitations'.

In particular: (a) in discussing various imaginary or real agents ('Alice' and 'Bob', 'Smith' and 'Jones', you and me) I leave completely open their personality, morality, gait and dress sense except where these matter to a decision. I'm not imagining persons who don't *have* personalities, gaits etc., but we needn't care what they are.

(b) There are three main idealizations. (b1) An agent is supposed to know the parameters of her decision situation. She knows: just what acts are available to her (e.g. virtue/sin); what the possible states of the world are (she is/is not predestined for salvation), though *not* which one obtains; and which outcomes are possible (she is saved/damned), though again not which one is actual. (b2) The agent is supposed to be ideally rational in the sense of (i) never making logical or mathematical mistakes, and (ii) knowing all relevant mathematical and logical truths. (b3) The agent is supposed to have an unrealistic fineness of psychological structure, in particular to have more preferences than any real person.

The simplification (a) and the idealization (b1) reflect the focus of this inquiry: practical reasoning in decision situations. It is not an inquiry into anyone's personality, morality etc., nor is it about *how* an agent knows that she is in this or that situation.

The idealization (b2) is connected with the inquiry's being normative, i.e. about what you *rationally ought* to do. In particular, it asks whether Evidential Decision Theory or Causal Decision Theory makes the better recommendation where they disagree. But 'making the better recommendation' had better mean 'making it *to otherwise rational agents*' if we are to find in this debate the significance that I claimed for it. It wouldn't matter much that CDT makes better recommendations than EDT to the logically deficient. The moral of *that* story would not be that causal knowledge is necessary for rational deliberation, but only that logical myopia can make it look that way.

Logical omniscience is consistent with the factual ignorance that the usually imaginary agents of the model share with real ones. *That* sort of ignorance is necessary for most decision problems to arise in the first place. It is hardly a decision *problem* whether to take an umbrella if you can already see that it is raining. Idealizing away that human limitation would leave nothing to discuss.

The idealization (b3) might be called a structural idealization, its point being just mathematical smoothness. Its precise content will soon become clearer, but whatever that is, the claim that agents have the superfine preferences that it demands is almost certainly false. It would probably be possible to frame the debate that concerns me, and all of the examples that follow, without it. But that would make for frequent, awkward and distracting qualifications. In any case, it is common ground to both EDT and CDT and so may reasonably be assumed when adjudicating between them.

The main task of this and the next chapter is to state EDT and CDT with more formality and precision than in the Introduction. Rather than doing this directly, I introduce them as alternative extensions of Savage's classic development.[1] Savage's work remains the most popular exposition of decision theory and so constitutes an accessible shop front. And it sharpens the contrast between EDT and CDT to present them as incompatible answers to a question that Savage leaves open.

1.2   States, events, outcomes and acts

It's natural to treat decision problems as having three types of element as distinguished by their places in Table 1.1. Here is what the table means: you must choose whether to take an umbrella on your walk to work this morning. It looks as though it might rain, but it's hard to say. If you take an umbrella then you won't get wet, but it is inconvenient having to carry one around. If you don't take one then it may turn out dry; but it may rain, and you will get to work wet, that being the worst possible conclusion of this unentertaining little drama.

---

[1] Savage 1972.

Table 1.1 *Weather Forecast*

	E_1: rain	E_2: no rain
O_1: take umbrella	Dry, carry umbrella	Dry, carry umbrella
O_2: don't take umbrella	Wet, no umbrella	Dry, no umbrella

The three element-types are as follows. First, there are two possible *states of the world*: that it rains and that it doesn't rain. These are arranged along the top row of Table 1.1. Second, there are two possible *acts*: that you take an umbrella and that you don't. These are listed down the far left-hand column of Table 1.1. Finally, each act-state pair determines a possible *outcome*. I mention each outcome in the body of the table in a place determined by the act-state pair that brings it about. They represent what the agent ultimately cares about; that is, what happens to him.

Savage's model retains this division but refines it considerably. First, **states** are highly detailed specifications of what goes on: they settle everything as yet unknown about the world.[2] We may identify them with *possible worlds* (which we can treat as points in a set rather than as concrete entities). Thus in Savage's model of *Weather Forecast*, there is a state of the world in which *it rains today and a Democrat is elected president in November 2052*. And there is another state of the world in which *it rains today and a Republican is elected president in November 2052*.

I'll call the set of all states the **state space S**. Any subset E of S (a set of possible worlds) is an **event**. So we can think of the top row of Table 1.1 as describing events. They are the ways that the world might be in so far as this makes a difference to the agent. E.g. the event 'Rain' is a set of possible worlds whose elements include every possible world at which it now rains and a Democrat wins in 2052, *and* all those at which it now rains and a Republican wins.

Second: the **outcomes** are what Savage calls 'consequences', i.e. 'anything that may happen' to the agent.[3] We can think of these again as possibilities or sets of possible worlds. But outcomes may be distinguished more finely than events: possible worlds in the same event may belong to different outcomes if the agent acts differently at them. Thus the outcome of *Weather Forecast* depends not only on whether it rains but also on whether you take your umbrella. I'll write Z for the set of possible outcomes.

---

[2] Savage 1972: 8–10.   [3] Savage 1972: 13.

Finally, the agent's available acts specify a particular outcome for each state. So we identify acts associated with different physical realizations. For instance, you may have a black umbrella and a blue umbrella. But since you don't care about the colour of any umbrella, but only whether you have to carry one around and whether you get wet, your taking the black umbrella and your taking the blue one get lumped together in the same act: that act such that, if you perform it and it rains then you are dry and carry an umbrella around, and if you perform it and it doesn't rain then again you are dry and carry an umbrella around.

More generally, Savage identifies **acts** with *functions* from the set of states to the set of outcomes. For instance the act *Don't take umbrella* is the function that takes any state in *Rain* to the outcome *Wet, no umbrella* and any state in *No rain* to the outcome *Dry, no umbrella*. Savage identifies the set A of acts with Z^S, the set of *all* functions from the states in S to the outcomes in Z.[4]

$A = Z^S$ means that Savage's 'acts' far outrun the options that are really available to any agent. Thus if in *Weather Forecast* we take *Rain* and *No rain* to be the *states* of the decision problem, then there are seven such acts in addition to the two actually available options that Table 1 mentions. Those nine acts are (in an obvious notation):

(1.1)   Rain → Dry, carry umbrella; No rain → Dry, carry umbrella
(1.2)   Rain → Dry, carry umbrella; No rain → Wet, no umbrella
(1.3)   Rain → Dry, carry umbrella; No rain → Dry, no umbrella
(1.4)   Rain → Wet, no umbrella; No rain → Dry, carry umbrella
(1.5)   Rain → Wet, no umbrella; No rain → Wet, no umbrella
(1.6)   Rain → Wet, no umbrella; No rain → Dry, no umbrella
(1.7)   Rain → Dry, no umbrella; No rain → Dry, carry umbrella
(1.8)   Rain → Dry, no umbrella; No rain → Wet, no umbrella
(1.9)   Rain → Dry, no umbrella; No rain → Dry, no umbrella

Of these acts only (1.1) and (1.6) correspond to anything really available. It isn't in your power e.g. to bring about (1.9), on which you are dry and have no umbrella whatever the weather, at least not if you are already committed to stepping outside.

This artificiality won't matter. As we'll see, Savage's theory constrains a rational agent's preferences amongst *all* acts in the present inflated sense. So it certainly constrains rational preferences amongst those acts that are really available. In this chapter 'acts' covers arbitrary elements of $A = Z^S$. The acts that are really available to the agent are his **options**.

---

[4] Savage 1972: 14.

1.3   Rational preference and the Savage axioms

Savage takes his agent to *prefer* some acts to others. Informally, if f and g are acts then his preferring f to g means that 'if he were required to decide between f and g, no other acts being available, he would decide on f'.[5] This is something that we might sometimes test empirically by observing the agent's choices. Alternatively, and to Savage's distaste, the agent might test his own preferences through introspection.[6] I'll write $f \succ_s g$ to say that the agent prefers f to g. Also I'll write $f \succeq_s g$ to say that the agent does *not* prefer g to f (f is 'at least as good as' g) and $f \sim_s g$ to say that the agent prefers neither to the other.

The content of the theory is that a rational agent's preferences conform to certain axioms. I state those axioms here in full. But nothing in what follows depends on their detailed content except for the three that I discuss below.

The Savage axioms depend on the following definitions. If f and g are acts and $E \subseteq S$ then $f = g$ *on* E means that $f(s) = g(s)$ if $s \in E$. $f \succ_s g$ *given* E means that $f^* \succ_s g^*$ for any f^*, g^* such that $f^* = f$ on E and $g^* = g$ on E and $f^* = g^*$ on $S - E$. E *is null* means that for any acts f and g, $f \sim_s g$ given E (that is, it is not the case that $f \succ_s g$ given E or that $g \succ_s f$ given E). For any $z \in Z$, a_z is the constant act such that $a_z(s) = z$ for any $s \in S$. If $z, z^* \in Z$ are outcomes then $z \succ_s z^*$ means: $a_z \succ_s a_{z^*}$. Finally, a *partition of* S is a set P of non-empty subsets of S such that every element of S is an element of exactly one element of P.

The axioms: for any events E, J, acts f, f^*, g, g^*, h and outcomes x, x^*, y, y^*, z:

(1.10)   If $f \succ_s g$ then $f \succeq_s g$; if $f \succeq_s g \succeq_s h$ then $f \succeq_s h$

(1.11)   For some outcomes v, w, $v \succ_s w$

(1.12)   Suppose $f = f^*$ on E, $g = g^*$ on E, $f = g$ on $S - E$ and $f^* = g^*$ on $S - E$. Then $f \succ_s g$ iff $f^* \succ_s g^*$

(1.13)   If E is not null, $f = a_x$, $g = a_y$ on E, then $f \succ_s g$ given E iff $x \succ_s y$

(1.14)   Suppose $f = a_x$ and $f^* = a_{x^*}$ on E, $f = a_y$ and $f^* = a_{y^*}$ on $S - E$; also $g = a_x$ and $g^* = a_{x^*}$ on J, $g = a_y$ and $g^* = a_{y^*}$ on $S - J$. And suppose $x \succ_s y$ and $x^* \succ_s y^*$. Then $f \succ_s g$ iff $f^* \succ_s g^*$

(1.15)   If $f \succ_s a_{g(s)}$ given E for all $s \in E$ then $f \succeq_s g$ given E. And if $a_{g(s)} \succ_s f$ given E for all $s \in E$ then $g \succeq_s f$ given E

---

[5] Savage 1972: 17. Savage himself faults this counterfactual criterion on the grounds that the agent might choose f and not g *faute de mieux* even if he were *indifferent* between f and g, were he offered a choice between them and nothing else. But this point undermines the criterion only if 'An indifferent agent might, if he were offered the choice between them, take f and not g' entails 'An indifferent agent would, if he were offered the choice between them, take f and not g'. The inference fails on one standard analysis of these counterfactuals (Lewis 1973b: 21–4).

[6] Savage 1972: 17. For an example of the introspective approach see Kreps 1988: 82–8. Sen 1977 defends this and other non-behaviouristic elicitations of an agent's preferences.

(1.16)   If $f \succ_s g$ then there is a finite partition P of S such that for every D $\in P$: (i) if $f^* = a_z$ on D and $f^* = f$ on $S - D$ then $f^* \succ g$; (ii) if $g^* = a_z$ on D and $g^* = g$ on $S - D$ then $f \succ_s g^*$.

The detailed content of these definitions and axioms isn't relevant here.[7] But a brief discussion of one definition and three axioms *is* in place. The definition of constant acts a_z etc. in effect extends preferences from acts to outcomes, since the act a_z is the act that gives outcome z in *any* state. Typically these 'acts' will not be among the options. For instance, in *Weather Forecast* neither of the constant acts (1.5) and (1.9) is available to you, although the constant act (1.1) is. But your preferences amongst them represent your preferences amongst the outcomes: for instance, that you prefer (1.1) to (1.5) is a way of saying in Savage's idiom that you prefer *Dry, carry umbrella* to *Wet, no umbrella*.

The first part of (1.10) says that if the rational agent prefers (say) *taking an umbrella* to *not taking one*, then he doesn't prefer *not taking an umbrella* to *taking one*. The second part says that the 'at least as good as' relation is transitive. That is, if the rational agent wouldn't always take h over g, and wouldn't always take g over f, then he wouldn't always take h over f. Whilst it may not be possible to justify these principles in more basic terms, it is at least difficult to question the first without rejecting the idea of rational preference altogether. Together they make preference transitive: if $f \succ_s g$ and $g \succ_s h$ then $f \succ_s h$.[8]

Informally, (1.12) says that a burglar who would rather take the diamonds in the safe than the cash in the attic should have this preference *whatever* prison sentence burglary attracts. More generally, it says that rational preference for one act over another depends *only* on the agent's preferences in the event that they have different outcomes. Whether the burglar takes the diamonds or the cash only makes a difference to him if he isn't caught.

That looks innocuous enough. But it will matter that it and (1.10) together entail the following (simplified):

(1.17)   ***Principle of dominance***: If P is a finite partition of S and if $f \succeq_s g$ given E for each $E \in P$, and if f and g are both constant on any such E, then $f \succeq_s g$. If, in addition, $f \succ_s g$ given some $E^* \in P$ (possibly $E = E^*$), then $f \succ_s g$.

---

[7] The axioms and definitions jointly have the same content as the postulates in the endpapers of Savage 1972. But they differ over the distribution of that content, on which point this account is closer to Kreps 1988: Chapter 9.

[8] Proof: if $f \succ_s g$ and $g \succ_s h$ then $f \succeq_s g$ and $g \succeq_s h$ by the asymmetry of $\succ_s$. So if $h \succeq_s f$ then $h \succeq_s g$ by the transitivity of $\succeq_s$. But this contradicts $g \succ_s h$. So $\neg h \succeq_s f$ i.e. $f \succ_s h$.

Table 1.2 Savage 2.7

	Democrat	Republican
Invest in farmland	$350	$200
Invest in gold	$250	$100

If the conditions in the first part of (1.17) are met then *f weakly dominates g with respect to P*. If in addition $f \succ_s g$ given some $E^* \in P$, then *f strictly dominates g with respect to P*.

For instance, suppose you prefer more money to less, that this is all you care about, and that your options are investing in farmland and investing in gold. You ask yourself what profit you would expect in each case, first assuming that the next president is a Democrat and then assuming that the next president is a Republican. Somehow you come up with precise answers to these questions (see Table 1.2).

In either event (*Democrat* or *Republican*) you are better off in farmland. That is, investing in farmland strictly dominates investing in gold with respect to this partition. The principle therefore prefers investing in farmland to investing in gold. This instance of it looks plausible enough, but we'll see that other cases make trouble for Savage.[9]

Informally, (1.16) says that if an act *f* is better than (i.e. preferred by the rational agent to) an act *g*, then however bad is outcome *z*, some region of the state space is so 'small' that getting *z* there, and *f* everywhere else in *S*, is *still* better than *g*. For instance, if in *Savage 2.7* you rationally prefer (*f*) investing in farmland to (*g*) investing in gold, then consider the act f^* of investing in farmland *unless* this fair coin lands heads on its next twenty tosses, in which case you invest in pork bellies for a fixed and guaranteed return of $100. Here the 'small' region of the state space is the set of possible worlds at which the coin does land heads on the next twenty tosses. And your rational preference for this f^* over *g* witnesses the truth of (1.16). That axiom hardly shares the intuitive force of the other two that I mentioned. But it is crucial to Savage's subsequent formal treatment,

---

[9] This example adapts one from Savage 1972: 21. The principle of dominance is an instance of Savage's 'Theorem 2' (1972: 24). To see why dominance follows from (1.10) and (1.12), let $\{E_1, \ldots E_n\}$ be a partition of S satisfying the conditions of (1.17). Then $f(E_i) \succeq_s g(E_i)$ for $i = 1, \ldots n$. For $j, k = 1, \ldots n$ let $h_j(E_k) =_{\text{def.}} g(E_k)$ if $k \leq j$ and $h_j(E_k) =_{\text{def.}} f(E_k)$ otherwise. So by (1.12) $f \succeq_s h_1 \succeq_s h_2 \ldots \succeq_s h_n = g$. So by (1.10) $f \succeq_s g$. Also if, for some $m, f(E_m) \succ_s g(E_m)$ then either $m = 1$ and $f \succ_s h_1$ or $m > 1$ and $h_{m-1} \succ_s h_m$. In either case, by (1.10) $f \succ_s g$.

as we'll see at section 1.4. It is an example of structural idealization as discussed at section 1.1(b3).

The point that we have reached is that the rational agent's preferences accord with the axioms (1.10)–(1.16). Savage now shows that we can make sense of the agent's attaching this or that numerical *probability* to the events in the decision situation that he faces. In other words, given enough facts about what the agent would *do* when faced with this or that choice, we can extract further facts about how likely the agent *thinks* are the events that determine the outcomes of his acts. Furthermore, we can also make sense of the idea that the agent attaches this or that numerical *utility* to the outcomes of that decision situation. And finally, the relation between preference, utility and probability is very simple: rational preference holds between f and g if and only if the sum of the utilities of each possible outcome arising from f, weighted by the probability of the events that f takes to it, exceeds the sum of the utilities of each possible outcome arising from g, weighted by the probability of the events that g takes to *it*.

This is all programmatic, and I'll cover these three steps in more detail in the next two sections. But it's worth briefly stating the relationship between this development and the overall project. *If* Savage's axioms tell the truth about rational preference then a rational agent always maximizes expected utility, in a weighted-by-probability-of-events sense that is common ground, at least nominally, between EDT and CDT. And we can then characterize the disagreement between *them* as a dispute over what sorts of things get to count as the events whose probabilities play this role.

1.4   From preference to probability

Intuitively, the thought that an event is 'probable', or 'more probable' than another, might have something to do with the frequency with which other events of the same type do actually occur. In that sense it would be true to say e.g. that snow this December is more probable (in Britain) than snow next July, because snow is more frequent in December than in July. If that *is* true, then it is true 'for' all of us. So actual frequency might be called an *impersonal* probability. Similarly, the *chance* of an event is an objective property of a possible event: for instance, a fair coin has the same chance of landing heads on its next toss as of landing tails on its next toss. Chance is impersonal. What I just said about that coin is true even if everyone thinks it is biased towards heads.

But we can extract a different type of probability, personal probability, from the things that people do. Intuitively your *personal* probability for a proposition, which we might also call your confidence in it, is reflected in your behaviour: in your choices amongst acts whose relative success depends on whether it obtains. Your taking an umbrella in *Weather Forecast* reflects a higher confidence in rain than does your not taking one. And, as attested by the fact that some people take umbrellas when their companions don't, the (personal) probability of rain will vary across persons in the same place at the same time.

Savage gives a thorough treatment of personal probability of which I present only those highlights that matter here. The basic idea (due to Ramsey[10]) is that if the agent prefers an act that wins a prize if it is *sunny* and invites a penalty if it is not, to an act that wins the same prize if it *rains* and invites the same penalty if it does not, then the agent considers sunshine more probable than rain. More precisely and more generally:

(1.18)   If $E, H \subseteq S$ then **E is more likely than H according to $\succ_s$**, written **$E > H$**, iff, for any $x, y \in Z$ s.t. $x \succ_s y$, if $f = a_x$ on E and $f = a_y$ on $S - E$, and $g = a_x$ on H and $g = a_y$ on $S - H$, then $f \succ_s g$.

Savage's axioms (1.10)–(1.16) imply that exactly one numerical function Pr_S of a particular form takes subsets of S to real numbers between 0 and 1, assigning a higher or lower number to any subset of S depending on its position in the > ranking. More formally:

(1.19)   There is exactly one function Pr_S: $\wp(S) \rightarrow [0, 1]$ with the following properties:
          (a)   **Pr_S represents >**: $Pr_S (E) > Pr_S (H)$ iff $E > H$
          (b)   **Pr_S is a probability function**:
                 (i) $Pr_S (E) \geq 0$ for any $E \subseteq S$
                 (ii) $Pr_S (S) = 1$
                 (iii) If $E \cap H = \varnothing$ then $Pr_S (E \cup H) = Pr_S (E) + Pr_S (H)$

The proof of this statement makes essential use of the idealizing axiom (1.16).[11]

The axioms (1.19)(b)(i)–(iii) are definitive of probability. I shall be using them repeatedly in the course of what follows. I shall also be referring to **conditional probability** defined as follows: the conditional probability of E given H is **$Pr_S (E|H)$** $=_{\text{def.}} Pr_S (E \cap H)/Pr_S (H)$ when $Pr_S (H) > 0$ and

---

[10] Ramsey 1990 [1926].

[11] For instance, in the case that S has five elements, there is some $\succ_s$ such that the corresponding > is not representable by any probability function, i.e. there is no such Pr_S satisfying (1.19)(a) (Kraft, Pratt and Seidenberg 1959). It takes (1.16) to rule this out. For the proof of (1.19) see Savage 1972: Chapter 3.

undefined if $Pr_S(H) = 0$. Intuitively $Pr_S(E|H)$ represents the proportion of the H-region of the state space that the E-region occupies. More informally still, we can think of it as representing the agent's confidence *that E obtains if H does*. I'll discuss conditional probability further in Chapter 2.

Three philosophical points of clarification are appropriate here. The first is simply to repeat that the probability function Pr_S as jointly defined by (1.18) and (1.19) is a *personal* probability. It depends on the preference function of the agent, so varies across rational agents. Since there is no one 'right' pattern of preferences amongst acts, there is no one 'right' personal probability function either. For all that I have said about it, personal probability neither reflects nor tries to reflect any feature of the common objective world that all of these agents inhabit.

The second point is that we shouldn't think of personal probabilities thus defined as psychologically internal states that *cause* the agent to have preferences over acts, of which, however, they are *logically* independent. Of course that *is* how we normally think of subjective confidence. For instance, your confidence that it will rain is something that you might cite in *explanation* of your preferring O_1 to O_2 in *Weather Forecast*. It serves as an explanation because it describes a mental state that would have been that same state even if you had had different preferences over these acts, and perhaps even if you had had no preferences at all.

But in the Savage construction, personal probability is not a piece of mental furniture that would be mentally as it actually is even if you behaved differently. It *is* just a complicated amalgam of preferences over acts. To speak of the agent's personal probability function just *is* to speak of her preferences over acts, and so not to identify anything that might explain them, just as saying that a picture is of a smiling face needn't explain the distribution of colours on the canvas but may just describe it in terms that hold special interest for us.

Finally, this conception of personal probability has obvious affinities with the behaviourist philosophy of mind, according to which one's mental states are no more than facts about one's actual or hypothetical behaviour. In particular, if (a) the agent's degrees of confidence in propositions (subsets of S) are Savage's personal probabilities, and if (b) the rational preferences on which these supervene are themselves simply facts about actual or counterfactual behaviour, then behaviourism about belief (= confidence) is true, at least for the rational agents of this model.

That conclusion is clearly attractive to anyone who (like the present author) feels sympathy for the positivist ideology of Savage's day. The mind is no unobservable mystery, but simply a construction from what is at least manifestly observable, if not always observed, viz the subject's actual and possible choices from amongst pairs drawn from A. Unfortunately and as we'll see, it turns out that (b) is untenable. So Savage's axioms cannot get us this far. But we can only see that once we have at hand the rest of the theory, to the brief exposition of which I now turn.

1.5   Utility and the representation theorem

That you prefer outcome x to outcome y carries only ordinal information, i.e. it concerns only your *ranking* of outcomes. It doesn't say *by how much* you prefer x to y. So to say that you prefer x to y and z to w is not yet to say whether you take the difference between x and y to be greater than, equal to or less than the difference between z and w. Is it possible to make a sense of these further claims by which we might evaluate them?

The answer is: yes, Savage's theory *gives* them that sense.[12] Informally, the idea is that we must distribute numerical values (or utilities) amongst outcomes in such a way that rational preference ranks more highly those acts that you *expect* to have greater utility according to Pr_S.

The following restricted form of Savage's result presents what is essential from the present perspective. For any $f \in A$, call a partition P of S an *f-partition* if for any $E \in P$, for any $s_1, s_2 \in E$, $f(s_1) = f(s_2)$, in which case $f(E)$ is well defined as the common value of $f(s)$ for any $s \in E$. Then:

(1.20)   If Pr_S is the probability function that represents $>$ then there is a real-valued **utility function** $U_S \colon Z \to R$, unique up to positive affine transformation, such that for any $f, g \in A$, if $P = \{E_1, \ldots E_n\}$ is a finite f-partition *and* g-partition of S then:

$$f \succ_s g \text{ iff } \Sigma_{1 \le i \le n} Pr_S(E_i)U_S(f(E_i)) > \Sigma_{1 \le i \le n} Pr_S(E_i)U_S(g(E_i))$$

Let us call an act *simple* if its outcome depends only on which of finitely many regions of the state space is actualized. Then what (1.20) says is that we can distribute utility over simple acts in such a way that $\succ_s$ prefers ones that have higher expected utility according to Pr_S. Equivalently, it says that the rational agent *maximizes expected utility* amongst simple acts: given a choice between two such acts, he chooses the one that maximizes the sum of the utilities of the outcomes that the act gives to each region, weighted by the confidence that Pr_S allocates to that region.

---

[12]   Savage 1972: Chapter 5.

Table 1.3 *Weather Forecast:* utilities

	E_1: rain	E_2: no rain
O_1: take umbrella	1	1
O_2: don't take umbrella	0	2

The restriction to *simple* acts is what allows us to speak of weighted *sums* (the Σ expressions in (1.20)). In many decision problems the acts are not simple. For instance, in a simple model of the decision to invest in a stock, the outcome of investment (one's real net profit) might be a decreasing function of a continuous variable (e.g. the rate of inflation), in which case investing is not a simple act. In these cases an analogous result holds in which appropriate integrals replace those weighted sums. But the acts that matter in what follows *are* simple. So it does no harm to focus on results that apply only to them. (The restriction to simple acts does not require that either S or Z be finite.)

To illustrate: note that by (1.20) we can (and I will) represent decision problems not by means of act/event/consequence tables such as Tables 1.1 and 1.2, but rather by means of act/event/utility tables. For instance, since the options in *Weather Forecast* are simple acts, we can tabulate a possible utility function U_S as in Table 1.3. What this says is that U_S (Dry, carry umbrella) = 1, that U_S (Wet, no umbrella) = 0 and that U_S (Dry, no umbrella) = 2. And what (1.20) tells us is that given this utility function on the outcomes we must have:

(1.21) $O_1 \succ_s O_2$ iff $Pr_S(E_1) + Pr_S(E_2) > 2\, Pr_S(E_2)$

Or equivalently, $O_1 \succ O_2$ iff $Pr_S(E_1) > Pr_S(E_2)$: you prefer taking the umbrella if and only if you are more confident that it will rain than that it won't.

It is a straightforward matter to extend the utility measure from states to simple acts:

(1.22)   If a probability function Pr_S and a utility function U_S represent an agent's preferences $\succ_s$ in the sense of (1.19) and (1.20), then the **Subjective Expected Utility (SEU)** of a simple act $f \in A$ relative to Pr_S, U_S is given by:

$$\text{SEU}(f) =_{\text{def.}} \Sigma_{1 \le i \le n} Pr_S(E_i) U_S(f(E_i))$$

where $\{E_1, \ldots E_n\}$ is an f-partition of S.

And it is a straightforward consequence of (1.20) and (1.22) that $f \succ_s g$ if and only if SEU (f) > SEU (g) relative to any Pr_S, U_S that represent $\succ_s$.[13] More concisely: Savage identifies practical rationality with *SEU-maximization*.

Principles (1.10)–(1.16), (1.19), (1.20) and (1.22) are the formal apotheosis of behaviourism. They construct *both* belief-like states (personal probability) *and* desire-like states (utility function) upon a mere preference ranking over acts. It is surprising and impressive that one can extract so much mental richness from such an exiguous basis. Perhaps it is not *so* surprising given the artificial latitude in Savage's notion of 'act', as remarked at section 1.2.

But the behaviouristic comments that I made about personal probability at section 1.4 do also apply to utility. We have no preference-independent way to identify an agent's utilities and probabilities. So we shouldn't, for instance, think of utility as directly measuring intensity of pleasure, or amounts of money, or anything else at which a rational agent might aim. Nor should we think of utilities and probabilities as jointly *causing* or otherwise *explaining* what the agent does.

Nor should we think of SEU-maximization as an internalizable principle that might guide deliberation. It is not as though, prior to making any decisions about what to do, the agent can first identify his utilities (say, as specified in Table 1.3), and his probabilities (say, 0.6 for *Rain*), and *then* distribute preferences across acts in accordance with SEU-maximization. Rather, his own probabilities and utilities are not identifiable as such, even to him, prior to the formation of any practical dispositions. Crudely: if he takes the umbrella then he thinks rain likely; but if he doesn't then that is not a mistake, because if he doesn't then he already thinks that rain is unlikely. More generally, *any* pattern of preferences amongst acts that satisfies the axioms is rational (=SEU-maximizing) relative to *some* probability function Pr_S and utility function U_S. That is the content of (1.19), (1.20), and (1.22); that is why their conjunction is standardly called a 'representation theorem'.

But this doesn't mean that the Savage axioms lack all normative content. On the contrary, they place quite strong demands on rational preference amongst acts. Some patterns of preference amongst acts are ruled out

---

[13] It is also easy to see that SEU is well defined, in the sense that (1.22) returns the same value for SEU (f) relative to a fixed Pr_S and U_S for any finite f-partition of the event space S. And it is easy to see that whilst SEU (f) will vary depending on which positive affine transformation U_S^* of U_S one chooses, any such choice will retain the ordering, i.e. SEU (f) > SEU (g) relative to Pr_S, U_S iff SEU (f) > SEU (g) relative to Pr_S, U_S^*.

as irrational. For instance, and as we saw at section 1.3, it follows from axiom (1.10) that if you prefer f to g and g to h, you cannot rationally fail to prefer f to h. So if we are given *some* preferences amongst acts, other preferences amongst other acts become rationally mandatory.

Thus, e.g. in *Weather Forecast*, if you (a) prefer taking the umbrella to the act (1.8) and (b) also prefer (1.8) to not taking the umbrella, then (c) you will, if you are rational, take the umbrella. It might even happen – though in this case it is unrealistic – that you determine the truth of (a) and (b) in advance of settling (c); in that case the Savage axioms offer guidance to practical deliberation. The point is that they only do this if you have *already* attained some preferences over acts by other means.

Considered as a normative principle, (1.10) is certainly intuitive: if you were to choose f over g and g over h when presented with these pairwise choices then it is certainly plausible that you should choose f over h in a straight choice between *them*. (It is another question whether any 'money pump' argument could constitute an economic justification of that principle.[14]) But taken together the axioms also have more questionable normative content, as I now discuss.

---

[14] Binmore 2009: 13–14 illustrates the argument as follows. Suppose that you (i) prefer apples to figs, (ii) prefer figs to oranges but (iii) do not prefer apples to oranges. Starting with an apple, you would swap it for an orange by (iii); by (ii) you would pay a small amount, say one cent, to exchange the orange for a fig; by (i) you would then pay another cent to exchange the fig for an apple: so you are back where you started but two cents poorer. By repeating these offers a person who owned a fig and an orange could easily bankrupt you. But this informal statement of the argument is ambiguous over the exact objects of your preferences. If e.g. (ii) means that you prefer *eating* figs to *eating* oranges, etc., then you might not swap the orange for the fig, if you think (as you might) that there is a reasonable chance that you will swap the fig for something else in the near future. If it means that you prefer holding a fig at (some fixed future time) t to holding an orange at t, then again you might not pay to swap the orange for the fig because you think that you will only swap the fig for an apple between now and t. Finally, if it means that you prefer holding a fig now to holding an orange now, then you will make the swaps, but this is not in consequence of any violation of (1.10). 'Figs now' does not denote the same option in 'I prefer figs now to oranges now', said at the second stage, as it does in 'I prefer apples now to figs now' said at the third stage. I have recently seen two more sophisticated money pumps for cyclic preference patterns that evade this difficulty (Rabinowicz 2000: 138 ff.; Dougherty 2013). But there are cyclic (and so intransitive) preferences that avoid these money pumps too, and indeed seem immune to any sort of money pump. Thus consider a *lexicographic* preference $\succ$ over ordered pairs (w, F), where w specifies one's holding of dollars and F specifies one of the possible fruit holdings, which I'll label A, B and C. We may stipulate that $\succ$ is separable in the sense that for any real numbers m, n and any F, F' we have $(m, F) \succ (m, F')$ iff $(n, F) \succ (n, F')$; we may then write $F \succ F'$ for $\forall m\ (m, F) \succ (m, F')$. Now define:
(i) $(w, F) \succ (w', F')$ iff *either* $w > w'$ *or* $w = w'$ and $F \succ F'$
(ii) $A \succ B \succ C \succ A$
This relation $\succ$ is both intransitive and, it seems, immune to any sort of money pump, however sophisticated.

Table 1.4 *Sink or Swim*

	E_1: drowns	E_2: doesn't drown
f: learn to swim	z_1: cost, death	z_2: cost, survival
g: don't learn to swim	z_3: no cost, death	z_4: no cost, survival

1.6   Dominance and fatalism

We saw at section 1.3 that the axioms (1.10) and (1.12) entail the principle of dominance (1.17), of which the following is a simple consequence: if $f \succ_s$ g given E on which both are constant, and if $f \succ_s g$ given $S - E$ on which both are likewise constant, then $f \succ_s g$. How plausible is this?

Alice is about to take a cruise. Should she learn to swim? Learning to swim has costs in money and time, but drowning is worse. She only cares about her money and her life. Table 1.4 represents her options and the outcomes that matter to her. It follows from what I just said, and is anyway natural to suppose, that Alice's preferences over these outcomes are $z_4 \succ_s$ $z_2 \succ_s z_3 \succ_s z_1$. (Alice has $z_3 \succ_s z_1$ because she would like to spend money before her death even if it is imminent.) There is also nothing wrong with taking E_1 and E_2 as events (subsets of S) on each of which the acts f and g are both constant. And since $E_1 \cup E_2 = S$, we may write $E_2 = S - E_1$. Now since $z_3 \succ_s z_1$ we have $g \succ_s f$ given E_1; and since $z_4 \succ_s z_2$ we have $g \succ_s$ f given $E_2 = S - E_1$. It follows by the principle of dominance that $g \succ_s f$.

Here is the same argument in English. 'It is of no practical use for a sailor to learn to swim. For either she will die by drowning, or she will not. If she will not die by drowning, then so far as this goes her learning to swim will be a waste of time; if she will die by drowning, it will not save her.'[15] This *fatalist* argument could equally show the pointlessness of *any* act done for some further end. It is plainly absurd: but if it is, then so (it seems) is the principle of dominance.

It is easier to see what is wrong with the informal argument than to see what is wrong with the formal principle. Fatalism errs by ignoring the possibility that what you do might be relevant to the further end that you (certainly) either will or will not achieve. In this particular instance, the false premise is this: if she will not die by drowning, then so far as this goes her learning to swim will be a waste of time. It might be that Alice *won't* die of drowning and yet her learning to swim will *not* be a waste of

---

[15] Ayer 1963: 238. I have changed the sex of the sailor to fit my example.

Table 1.5 *Sink or Swim* reformulated

	$E_1: f \to D,$ $g \to D$	$E_2: f \to D,$ $g \to \neg D$	$E_3: f \to \neg D,$ $g \to D$	$E_4: f \to \neg D,$ $g \to \neg D$
f: learn to swim	z_1: cost, death	z_2: cost, death	z_3: cost, survival	z_4: cost, survival
g: don't learn to swim	z_5: no cost, death	z_6: no cost, survival	z_7: no cost, death	z_8: no cost, survival

time (or money), since it is *because* she learns to swim that she does not drown.

Returning to the formal principle of dominance: the mistake was to treat the E_1 and E_2 that I specified in Table 1.4 as appropriate subsets of the state space; that is, as events. The chosen partition of the state space must make *independent* of the agent's act which element of the partition the actual world is in, in some sense of 'independence' in which (say) whether Alice drowns is *not* independent of whether she learns to swim. For instance, if independence means '*causal* independence' then Table 1.4 is misleading in a way that Table 1.5 is not. In this table, 'D' and '¬D' describe Alice's drowning or not, and '$\to$' represents deterministic causation. The event E_1 (which might encompass e.g. piracy) is that nothing Alice now does can save her. The event E_2 is that learning to swim would cause Alice's death. This possibility is bizarre and unlikely. The event E_3 (perhaps encompassing shipwreck near a desert island) is that learning to swim would save her from drowning but not learning to swim would doom her to drowning. Finally, E_4 covers the most likely situation: that the journey is without incident.

Principle (1.17) does *not* apply to the partition implicit in this representation of the problem. For $g \succ_s f$ given E_1, E_2 or E_4, but $f \succ_s g$ given E_3. And if Alice's $Pr_S (E_2)$ is low, her $Pr_S (E_3)$ is relatively high, and her $U_S (z_1)$, $U_S (z_2)$, $U_S (z_5)$ and $U_S (z_7)$ are all very low, then (1.20) gives what is in that case the sensible result that it is rational for her to learn to swim.

What this example shows is not that (1.17) is completely wrong, but that in order to be sure that we are applying it correctly, we need a general principle for deciding what is and what is not a permissible partitioning of *S*. The basic idea was implicit in the informal refutation of fatalism, in particular in the point that Alice's future survival might depend on her now learning to swim. It is this: whether an element of the partition describes actuality should be in some sense *independent* of what the agent does, i.e. of her options. In what sense?

There are two natural answers: causal independence and evidential independence. As a first approximation, we can say that the events are *causally* independent of the options if no choice of option would *make* any difference to which event (which element of the partition) is actual. And the events are *evidentially* independent of the options if no choice of option is a *sign* that this rather than that event occurs, whether or not there are any causal relations between them.

Briefly to repeat a point from the Introduction (section 0.2(b)): causal and evidential independence are certainly different relations. Two events might be causally independent whilst being evidentially dependent. For instance, causes are causally independent of their own effects whilst still being evidentially dependent upon them. And joint effects of a common cause are both causally independent and symptomatic of one another. My receiving a Christmas card from my grandmother is a sign that my cousin in Australia has got one too, but these events are causally independent.

It is not so common to find a divergence between causal and evidential independence of an agent's options. For instance, it is plausible that which of $E_1 - E_4$ in Table 1.5 is actual is causally *and* evidentially independent of whether Alice learns to swim. However, such cases *are* at least possible. And they will be crucial in the argument to follow. The reason for this is that Evidential and Causal Decision Theory diverge over just this point. The Evidential Theory says in effect that SEU-maximization applies only to partitions that are *evidentially* independent of the agent's options. But the causal theory says that it applies to all partitions that are thus *causally* independent.

I said that this characterization of the two kinds of independence was a first approximation to what I had in mind. The second approximation makes trouble for Savage's behaviourist ambitions. To see what it is, note first that the point of normative *subjective* decision theory is to say what is rational *given the agent's beliefs and desires*. This is true even on Savage's construction, which reduces beliefs and desires to patterns of preferences over acts. The Savage approach places constraints on which such patterns of preferences are rationally permissible. And it *thereby* places constraints on which preferences over acts are rational given this or that combination of beliefs and desires.

For instance, if its known bias, or anything else, makes me almost certain that this coin will land tails then the subjective theory ought to make betting tails the rational choice, even if it is *in fact* true that this coin will land heads. What is in fact true should not enter into the calculation at all, for what interests us is a kind of rationality to which completely mistaken as well as relevantly well-informed persons can equally attain. (I defend this subjectivism at section 3.3 below.)

For this reason, the ultimate account of whether a partition is suitable for SEU-maximization must make that depend only on the agent's *beliefs* about whether the partition is independent of the available options. Thus in *Sink or Swim* the decision to learn to swim may be rational if Alice *thinks* that it makes no difference to which of $E_1 - E_4$ in Table 1.5 is actual. If learning to swim does in fact tend to make E_2 more likely, but there is no way that Alice could have known this, then her learning to swim is no more irrational than my betting tails on the toss of a (tails-biased) coin which will in fact land heads on this occasion.

So it seems that we must build in the following specification from the start: an appropriate partition of S into events must be one that the agent *believes* to be (somehow: either causally or evidentially) independent of which option he chooses. But then we must *already* have a conception of what it is for the agent to have such beliefs. For instance, the basic idea behind (1.18) is, roughly, that the agent is more confident that it will be sunny tomorrow than that it will rain tomorrow just in case he prefers betting on sunshine, at given stakes and odds, to betting on rain at those same stakes and odds. But now we see that that is wrong. What is right is this: that for the agent to think sunshine more probable than rain is for him to prefer the first bet to the second *if* he already *believes* that whether there is sunshine or rain tomorrow is (in some sense) independent of his choice between these bets.

So it turns out that the agent's beliefs are after all *not* just facts about his actual or possible behaviour. They are facts about the actual and possible behaviour of an agent equipped with such-and-such other beliefs, in particular with other beliefs about the independence of events from acts. So we must give up on the aim of such purely operational explications of beliefs and desires as Savage's personal probability and utilities were intended by him to be.

But once we have given up on *that* we have no good reason to retain the baroque and unintuitive structure of acts that Savage had to postulate. The reason is that as we'll see, there is an alternative, Evidential Decision Theory, that does away with it altogether. The theory is certainly not reductive: it does not identify an agent's beliefs and desires in terms that mention only her behaviour. But then Savage's construction cannot accommodate that ambition either.[16]

[16] Taking a hint from Table 1.5, one might think that we *can* stipulate a condition on the partition of the state space that (a) suffices for the application of SEU-maximization and (b) does not mention either independence or the subject's beliefs about it. The idea would be to identify events with functions from *options* (available acts) to outcomes, so that if O is the set of options then the partition is Z^O, each event specifying what *would* be the outcome of each available act. But since Savage's model requires $\succ_S$ to be defined over *all* functions from states to outcomes and not only over options, we must now take $A = Z^{Z^O}$, and it is not even clear what it *means* to say that the subject would choose one over another arbitrary element of A so defined, which makes something of a mockery of the positivism that animates the whole approach (Gilboa 2009: 114–16; see Joyce 1999: 118–19 for this and further criticism of a similar idea). Alternatively, we could *directly* associate a probability with each function from options to outcomes (see e.g. Jeffrey 1977). But as with the approach that stipulates (believed) independence, that requires a prior understanding of what the probabilities mean and so rules out the possibility of behaviouristic reduction.

CHAPTER 2

EDT and CDT

Dropping the behaviouristic foundations of Savage's construction needn't mean losing the whole thing. Two of its central features remain prominent in the two rival developments that this chapter outlines. First, we can still explain some elements of the agent's mental life in terms of others. We can explain subjective probability and utility in terms of a relation of preference, although the story is in one way slightly more complicated than Savage's (see section 2.3). The main difference is that preference itself gets no behaviouristic analysis. Second, both EDT and CDT treat practical rationality as the maximization of expected utility, i.e. as a weighted sum of the utilities of outcomes, where the weights are probabilities. Where they differ, from one another as well as from Savage, is over how to calculate these sums. This reflects their differing interpretations of the independence condition, as discussed at the end of the last chapter.

But non-behaviouristic decision theory can possess a normative force that Savage's theory simply lacks. For instance, it is now possible that your utilities for the outcomes in *Weather Forecast* are as in Table 1.3, that you are very confident that it won't rain, and yet still you *take* your umbrella. This is possible because we don't identify utility and degrees of belief with patterns of rational behaviour. Instead they are mental states that even an irrational actor might possess and in light of which we might criticize what he does. Pursuing the analogy from section 1.4: Savage's theory identifies the content of a picture with generic properties of the distribution of paint on canvas: nothing, on this view, could so much as *be* a *very* poor depiction of a smiling face. But non-behaviouristic theories allow the identity of an external subject to enter into this content. So, for instance, a really badly executed picture of Smith's face would nonetheless be a picture *of him* and subject to criticism as such. Of course this still leaves room for disagreement over the justice of this or that critical standard.

2.1   Preference over news items

The word 'Gospel' derives from an Anglo-Saxon word meaning 'good news'. Specifically, the English word denotes the news that Jesus' sacrifice atones for Man's sin. Whether that is either good or news, it at least makes sense to call it that. A piece of information may be good news for you whether or not it is true and whether or not you can do anything about it. It is good news for you to the extent that you'd welcome it.

You might similarly call one piece of information *better news* for you than another, whether or not either is true or up to you. It is, for example, not only meaningful but also correct to call it better news for me (i) that I will win a big prize in this lottery than (ii) that I will win a small prize. This is compatible with (i) and (ii) being both false and both beyond my control.

It is in the sense just illustrated that I shall call one news item *better news for a specified agent* than another. This relation of *preference over news items* is fundamental. From now on I'll write $P \succ_A Q$ to mean that P is better news than Q for agent A. (To avoid confusion with Savage's preference relation, I will never use 's' to denote an agent.) If Q is not better news for the agent than P I'll write: $P \succeq_A Q$. If neither is better news than the other I'll say that the agent is *indifferent* between P and Q: $P \sim_A Q$. If the identity of the agent is obvious I'll just write $P \succ Q$ etc.

The following explication of preference over news items covers three essential points. First, $\succ$ is subjective. Second, $\succ$ relates propositions. Third, the agent's actual choices need not reveal it.

By calling $\succ$ *subjective* I mean (a) that agents might have different preferences over news items, and (b) that there is no question of their being 'right' or 'wrong'. For instance, if Alice and Bob have the same *beliefs* on all matters but different *tastes* then we might have $P \succ_{\text{ALICE}} Q$ and $Q \succ_{\text{BOB}} P$, where P is the news that today's dessert is ice cream and Q is the news that it is trifle. (Clearly neither one is *right*.) This could also happen if they have different beliefs. If Alice and Bob like both puddings equally, but Alice has the theory that ice cream is more nutritious than trifle whereas Bob is unsure which is more nutritious, then we might have $P \succ_{\text{ALICE}} Q$ but $P \sim_{\text{BOB}} Q$.

I make $\succ$ relate propositions and not (say) states of affairs or objects so as to distinguish news items that subjects in their ignorance *will* distinguish. For instance, if P^* is the news that today's dessert is the more nutritious out of ice cream and trifle, and Q^* is the news that it is the less nutritious, then we may have $P^* \succ_{\text{ALICE}} Q^*$ and $P^* \succ_{\text{BOB}} Q^*$, since Alice and Bob both want nutrition. But since Alice is right that ice cream is more nutritious, the state of affairs that the more nutritious food is today's pudding *is* the state of affairs that ice cream is today's pudding. So identifying news items with states of affairs would make $P^* \succ_{BOB} Q^*$ incompatible with $P \sim_{BOB} Q$. It would be wrong, for similar reasons, to identify news items with the located physical entities that are often called objects of people's desires. 'Alice prefers ice cream to trifle' should be read as 'Alice prefers the proposition that there is ice cream for pudding to the proposition that there is trifle for pudding.'

By 'propositions' I of course mean objects of people's beliefs as well as of their desires. It is convenient to identify these with sets of possible worlds; this is also possible given the idealizing assumption (b2) in section 1.1 above. Note three consequences of this: (i) logically equivalent propositions are identical; (ii) there is only one logical truth, here labelled T; (iii) there is only one logical falsehood, here labelled $\bot$.

Finally, I am not assuming about $\succ$ what Savage assumed about $\succ_s$, namely that if $P \succ_{BOB} Q$ then Bob will, if offered a choice between P and Q, realize P rather than Q. More generally, consider the following two claims:

(2.1)   **Preference constrains choice:** If $P \succ_A Q$ then A never chooses Q (i.e. to make Q true) when P is available.

(2.2)   **Preference constrains rational choice:** If $P \succ_A Q$ then A never rationally chooses Q when P is available.

I won't be assuming *either* (2.1) *or* (2.2). In the case of (2.1), the point is not that the assumption is false but that the present enquiry, being normative rather than descriptive, has no need for it. Claim (2.2) is true from the perspective that the book defends. But to assume it at this point would beg the whole question. I am here arguing *for* and not *from* the claim that preference over news value is the measure of rational choice.

To set aside the revelation of preference by means of choice is to depart from the standard approach of modern economic theory. This doesn't reflect any disagreement with that approach but only the fact that 'preference' has a different content here from that suitable for such descriptive enquiries.[1] But it raises the question: what content *does* it have, to say that one prefers a news item P to a news item Q?

Properly answering this question would take us further afield than is compatible with a concise treatment. The short answer is that one can

---

[1]   For statements of (2.1) and (2.2) see resp. Varian 2003: 121 and Varian 1992: 98.

often tell by introspection whether or not one would prefer the news that
P to the news that Q. This gives each of us an acquaintance with the
relation. That is inadequate as it stands. But clearly there is *a* relation of
preference that makes sense to us quite apart from anyone's actual choices.
You might now prefer the news that you are of royal blood to the news
that you are of common descent, even though it hardly makes sense to
ask which of them you would *now* bring about if you could. In any case,
whatever that relation is, it is a common primitive in both versions of
decision theory that matter here, and so for purposes of comparing them
perhaps no more than this outline is necessary.

2.2   The Jeffrey–Bolker axioms

As with the Savage construction, we now assume an agent, Jones, to whom
the idealizations 1.1(b1)–(b3) apply. The agent has a preference relation
$\succ_{\text{JONES}}$ over propositions and corresponding relations $\succeq_{\text{JONES}}$ and $\sim_{\text{JONES}}$
(but usually I'll just write $\succ$ etc.).[2]

In this scheme the analogues of Savage's axioms (1.10)–(1.16) require the
following definitions and suppositions. Propositions are subsets of the set
S of possible worlds. A set B of propositions is a *Boolean algebra* if (i) T
and $\bot$ are elements of B; (ii) $P \cap Q$ (that is, $P \wedge Q$) and $P \cup Q$ (that
is, $P \vee Q$) are elements of B if P and Q are; (iii) $S - P$ (that is, $\neg P$) is
an element of B if P is. The *supremum* of a set C of propositions is a
proposition R such that $P \wedge R = P$ for every $P \in C$ and $R \wedge R' = R$
for every R' that has this property. So e.g. the supremum of $\{P, Q\}$ is
$P \vee Q$. The *infimum* of C is a proposition R such that $P \wedge R = R$
for every $P \in C$ and $R \wedge R' = R'$ for every R' that has this property. So
e.g. the infimum of $\{P, Q\}$ is $P \wedge Q$. A Boolean algebra B is *complete* if
the supremum and infimum of C are elements of B whenever $C \subseteq B$. An
atom of B is an $X \in B$ such that for any $P \in B$, $P \wedge X = P$ if and only if
$P = \bot$ or $P = X$. B is called *atomless* if $\bot$ is the only atom of B.

The axioms governing $\succ_{\text{JONES}}$ are, then, as follows: for any propositions
P, Q and R:

(2.3) $\succ$, $\succeq$ and $\sim$ are defined over a set $B - \{\bot\}$ of propositions, where
        B is a complete and atomless Boolean algebra.
(2.4)   If $P \succ Q$ then $\neg(Q \succ P)$; if $P \succeq Q$ and $Q \succeq R$ then $P \succeq R$.
(2.5)   If $P \wedge Q = \bot$ then (a) if $P \succ Q$ then $P \succ (P \vee Q) \succ Q$; (b) if P
        $\sim Q$ then $P \sim (P \vee Q) \sim Q$.

---

[2] The expository material in this section is largely drawn from Jeffrey 1983: Chapter 9. See also Bolker
     1967.

(2.6) If $P \land Q = \bot$, $P \sim Q$, then if $(P \lor R) \sim (Q \lor R)$ for some R such that $P \land R = Q \land R = \bot$ and $\neg(P \sim R)$, then $(P \lor R) \sim (Q \lor R)$ for every such R.

(2.7) Let $\pi = \{P_1, P_2 \ldots\}$ be a sequence of propositions such that $P_n \land P_{n+1} = P_n$ for each n (or $P_n \land P_{n+1} = P_{n+1}$ for each n). Then if P^* is the supremum (resp. infimum) of π and $Q \succ P^* \succ R$ then for some N, $Q \succ P_n \succ R$ for any $n > N$.

As with Savage's axioms, a detailed understanding of these claims is not necessary for what follows. But a few comments are in order. Axiom (2.4) is the analogue of (1.10) in Savage's scheme: it implies, for instance, that if you prefer the news that P to the news that Q and the news that Q to the news that R, then you prefer the news that P to the news that R. As before, the axiom is plausible. At any rate I shall not enter into further justification of it.

Axiom (2.5) says roughly that if one proposition is better news than another with which it is incompatible, then their disjunction is 'between' them. For instance, consider the proposition that either Jones or Smith has won first prize in some competition. Jones would welcome this *more* than the news that Smith has definitely won it but *less* than the news that he, Jones, has definitely won it.

Axioms (2.3), (2.6) and (2.7) are not plausible. Rather, they are idealizations of the structural type as described at 1.1(b3). For instance, atomlessness requires that $\succ$ be defined over a field in which there are no *maximally* 'fine' propositions (none corresponding to individual possible worlds) but that there are arbitrarily fine ones: e.g., the sequence of propositions P_1, $P_2, \ldots$ where P_n says that the mean temperature at place X at time T was between $25\,°C$ and $(25 + 1/n)\,°C$. My only excuse for these axioms is that they are necessary for the mathematical framework in which it is customary to pose the main questions here addressed. Doing without them would probably be possible but would certainly make things more awkward.

In any case, the point is that if Jones's preferences *do* respect (2.3)–(2.7), we can construct functions, analogous to Savage's Pr_S and U_S, that interact to represent the agent's preferences in a similar way. In particular:

(2.8) If $\succ$ satisfies (2.3)–(2.7) then there is a **value function** $V\colon B \to \mathbf{R}$ and a probability function $Pr\colon B \to [0, 1]$ such that:

    (i) V represents $\succ$ in the sense that for any $P, Q \in B$, $V(P) > V(Q)$ iff $P \succ Q$;

    (ii) For any $P \in B$ and any $\pi \subseteq B$ s.t. $\pi = \{S_1, S_2 \ldots S_n\}$ is a partition of S, $V(P) = \Sigma_{1 \le i \le n}\, Pr(S_i \mid P)\, V(P \land S_i)$[3]

---

[3] In the case that $Pr(P \land S_i) = 0$ we set the relevant summand $V(P \land S_i)\, Pr(S_i \mid P)$ to 0.

(2.8) says that we can always represent Jones's preferences for news items as a ranking that ***maximizes conditional expected utility***: that is, given any range of (exhaustive, exclusive) alternatives $S_1, \ldots S_n$, Jones ranks any proposition P according to the sum of its welcomeness to him in S_n, weighted by the probability of S_n *given that the proposition P is true*, this weighting being reflected in the appearance of conditional probabilities $Pr(S_i \,|\, P)$ on the right-hand side of (2.8)(ii).

The representation of $\succ$ at (2.8) differs in two obvious ways from the representation of $\succ_s$ at (1.19) and (1.20). Most obviously, (2.8) applies indifferently to all propositions and makes no distinctions between what the Savage formulation would distinguish as acts, events and consequences. What this means is that it simply does not matter *how* we partition S into 'events': any partition satisfying the conditions of (2.8)(ii) will return the same result, as far as V is concerned, as any other. This 'partition-invariance' constitutes a great simplification.

What is responsible for this invariance is a second difference from Savage: in (2.8)(ii) the probabilistic weights are *conditional* probabilities of the S_i upon P (cf. Savage: *un*conditional probabilities of these events). To see the difference that this makes, reconsider *Sink or Swim* (Table 1.4). Assume that the function Pr reflects the statistical facts that amongst those who cannot yet swim, 0.01% of those who *learn* to swim prior to a lengthy sea journey drown on that sea journey, whereas 1% of those who do not then learn to swim drown on the journey. What this means is that $Pr(E_1 \,|\, f) = 0.0001$ and $Pr(E_1 \,|\, g) = 0.01$. Since Pr is a probability distribution and $\{E_1, E_2\}$ a partition, we have $Pr(E_2 \,|\, f) = 0.9999$ and $Pr(E_2 \,|\, g) = 0.99$. (Here I am writing 'E_1', 'E_2', 'f' and 'g' for the propositions describing the occurrence or realization of the corresponding events or acts in Table 1.4.) Assuming that $\succ_{\text{ALICE}}$ satisfies (2.3)–(2.7), (2.8)(ii) then gives:

(2.9) $V(f) = 0.0001\, V(f \wedge E_1) + 0.9999\, V(f \wedge E_2)$

(2.10) $V(g) = 0.01\, V(g \wedge E_1) + 0.99\, V(g \wedge E_2)$

So there certainly are real-valued value functions V on B that give f greater value than g (from Alice's perspective). This will be so if any proposition that entails Alice's drowning on the voyage has very much lower value for her than any that does not. In other words, if Alice greatly values not drowning over drowning then she will value learning to swim over not learning to swim.

2.3   Credence

But (2.8) also differs from (1.19) and (1.20) in a less obvious way. Whereas the probability function Pr_S introduced at (1.19) was *unique*, we cannot

Table 2.1 *Jones and the Auto Rental*

	¬*L*: arrives on time	*L*: arrives late
A: rents Audi	$A \wedge \neg L$	$A \wedge L$
¬*A*: rents Škoda	$\neg A \wedge \neg L$	$\neg A \wedge L$

say the same of *Pr* as described at (2.8). In fact, unless ≻ meets one of certain additional conditions that arguably entail irrationality, there is a continuum of functions *Pr* satisfying (2.8)(ii).[4] Worse, these functions do not even agree *ordinally*. Typically there are Pr_1, Pr_2 satisfying (2.8)(ii) such that $Pr_1 (P) > Pr_1 (Q)$ but $Pr_2 (Q) \geq Pr_2 (P)$ for some propositions *P*, *Q*.[5] This indeterminacy in *Pr* also entails an indeterminacy in the value scale, although of course (2.8)(i) implies that any *V*, *V** that both satisfy (2.8) must at least agree ordinally with any other.

An artificially simple example of this would be Jones's situation as his flight approaches London Airport. On landing he must hire a rental car to reach Swindon in time for a meeting. An Audi costs more to hire than a Škoda but will get him there more quickly. Aside from *T* there are eight relevant propositions; these appear in the headings and the body of Table 2.1. It is consistent with this story and also with the axioms (2.3)–(2.7) that ≻JONES ranks these nine propositions as follows:

(2.11) $\neg A \wedge \neg L \succ \neg L \succ A \wedge \neg L \succ A \succ T \succ \neg A \succ \neg A \wedge L \succ L \succ A \wedge L$

Thus e.g. the news (*A*) that only Audis are available is more welcome to Jones than the news (¬*A*) that only Škodas are available. In this story, the preferences (2.11) fit a pair (*Pr*, *V*) reflecting the beliefs that the Audi is only slightly faster and only slightly costlier than the Škoda and on which Jones is as likely as not to be late. But they also fit a pair (*Pr**, *V**) according to which the Audi is very much faster and very much more expensive and on which Jones is about *half* as likely to be late as not. So ≻JONES does not determine a unique probability function. For the same reason, his choosing the Audi when the Škoda is available does not pin down his beliefs, of which (2.8) therefore cannot be said to be reductive in the way that (1.18) and (1.19) purported to be. More generally, there is no hope of identifying Jones's degrees of confidence with *whatever Pr* satisfies (2.8)

---

[4] Such a condition is Jeffrey's (8–17) (1983: 142), which implies that *V* is unbounded and thus arguably leads to irrational preferences in the St Petersburg game (Jeffrey 1983: 150–3).

[5] More specifically, if *Pr*, *V* satisfy (2.8) then so do $Pr^* = (cV + d) Pr$ and $V^* = (aV + b) / (cV + d)$ for any real numbers *a*, *b*, *c*, *d* such that $ad - bc > 0$, $cV + d > 0$ and $cV (T) + d = 1$ (Jeffrey 1983: 96–9).

given his preferences over news items, the candidate functions being too many and too various.[6]

Instead I propose to treat 'equal confidence' as primitive. That is, I am going to assume that we all know what is meant by saying that you are just as confident of the truth of some proposition P as of some other proposition Q, where these are both possible, and that we understand this independently of the preference relation. And I am going to assume that for any particular agent A, there are propositions P and Q such that the agent prefers P to Q (considered as news items) *and* is equally confident of both. For instance, let P be the proposition that the agent wins a large cash prize in some fair lottery and let Q be the proposition that some other specified person wins the same prize in that lottery. Finally, I will assume that amongst the many function-pairs (Pr, V) that satisfy (2.8) relative to a given agent's $\succ$, there is *at least* one such pair (Pr, V) such that Pr represents the agent's equal-confidence relation in the sense that $Pr(X) = Pr(Y) \neq 0$ if the agent is equally confident of X and Y where these are not impossible.

If all of these conditions are met then there is *exactly* one Pr such that (Pr, V) satisfies (2.8) and Pr represents the equal-confidence relation.[7] The corresponding V is determined up to positive affine transformation.

---

[6] For a specific example, the following table gives to two decimal places the relevant values of such function pairs (Pr, V) and (Pr^*, V^*):

	$\neg A \wedge \neg L$	$\neg L$	$A \wedge \neg L$	A	T	$\neg A$	$\neg A \wedge L$	L	$A \wedge L$
Pr	0.13	0.5	0.38	0.5	1	0.5	0.38	0.5	0.13
V	4	2.5	2	0.5	0	−0.5	−2	−2.5	−4
Pr^*	0.19	0.66	0.47	0.53	1	0.47	0.28	0.34	0.06
V^*	3	2.14	1.8	0.53	0	−0.6	−3	−4.09	−9

Note that $Pr(L) = Pr(\neg L)$ but $Pr^*(L) < Pr^*(\neg L)$. Of course the example is artificially simple because it ignores the requirement that $\succ_{\text{JONES}}$ be defined over an *atomless* field. Expanding the field to meet that condition needn't affect this illustration of the point.

[7] Suppose:
  (i) (Pr, V) and (Pr^*, V^*) satisfy (2.8) for a given agent's $\succ$
  (ii) Pr and Pr^* both represent that agent's equal-confidence relation
  (iii) P, Q are such that $V(P) > V(Q)$ and $Pr(P) = Pr(Q) \neq 0$
Then:
  (iv) $\exists a, b, c, d$ s.t. $ad - bc > 0$, $V^* = (aV + b) / (cV + d)$, $Pr^* = (cV + d)Pr$ and $cV(T) + d = 1$ by (i) and Jeffrey 1983: Chapter 8
  (v) $Pr^*(P) = Pr^*(Q)$ by (ii), (iii)
  (vi) $(cV(P) + d)Pr(P) = (cV(Q) + d)Pr(Q)$ by (iv), (v)
  (vii) $cV(P) + d = cV(Q) + d$ by (iii), (vi)
  (viii) $c = 0$ by (iii), (vii)
  (ix) $d = 1$ by (iv), (viii)
  (x) $Pr^* = Pr$ and $V^* = aV + b$, $a > 0$ by (i), (viii), (ix)

This probability function is the agent A's **credence function** Cr_A (which I'll usually write Cr). I shall take it to measure her strength of confidence in the propositions to which it applies. The corresponding V-function (or rather, the family of its positive affine transformations) is the agent's **news value function** V_A (which I'll usually write V). Since every agent in the examples that follow can, I think harmlessly, be supposed to meet these assumptions, we can represent the intensities of her beliefs and desires (in the sense of news values) by Cr and V respectively.

One further assumption that is common in the literature and which this book shares is that just as the probability function Cr reflects the agent's beliefs, so too the *conditional* probability functions that it induces reflect the agent's judgements of evidential bearing. More particularly, $Cr\,(X \mid Y)$ $=_{\text{def.}} Cr\,(X \wedge Y)/Cr\,(Y)$ reflects the extent to which the agent *would believe* X *on learning that* Y (certain pathological cases aside[8]). And $Cr\,(X \mid Y) -$ $Cr\,(X \mid \neg Y)$ reflects the degree of evidential support that the agent takes Y to give X, so that $Cr\,(X \mid Y) = Cr\,(X \mid \neg Y)$ iff the agent takes Y to be evidentially irrelevant to X.

Further discussion of this *Bayesian* assumption is largely beyond the scope of this book. But it is generally taken for granted in the dispute between EDT and CDT.[9] One reason it is convenient is that we can apply the machinery of (2.8)(ii) directly to agents about whom we are given only their views on evidential bearing, without first having to calculate or conjecture whatever unconditional probabilities sustain these views. Such agents figure largely in the main sort of application of the theory, to which I now turn.

2.4 Evidential Decision Theory

Perhaps the most obvious decision-theoretic problems in this framework would take the following form: given a specification of $\succ$ and the equal-confidence relation, deduce the agent's beliefs and desires, i.e. Pr and V. But another sort of problem, the sort that concerns us here, takes a different form: given a *partial* specification of an agent's beliefs and desires, calculate her news value for *other* propositions. Here we are being told something about what the agent believes and what she values as news; the problem is then to establish what *else* the agent so values.

---

[8] For instance, the case that e.g. X is the proposition that nobody believes Y. For discussion of a similar case see Mellor 1993: 243.

[9] For criticisms of Bayesianism that are independent of the issue over causalism see e.g. Levi 1974; Glymour 1980: Chapter 3.

For instance, we might be told the following about the agent's beliefs and news values in *Jones and the Auto Rental*:

(2.12) $Cr(L \mid A) = 0.25$

(2.13) $Cr(L \mid \neg A) = 0.75$

(2.14) $V(L \wedge A) = -4$

(2.15) $V(L \wedge \neg A) = -2$

(2.16) $V(\neg L \wedge A) = 2$

(2.17) $V(\neg L \wedge \neg A) = 4$

Propositions (2.12) and (2.13) tell us something about Jones's beliefs: his conditional probabilities of being late given that he does (does not) take the Audi, these being facts about his actual state of mind. In particular, we know that he is three times more confident of being late given that he takes a Škoda than of being late given that he takes an Audi. Propositions (2.14)–(2.17) tell us something about his beliefs: they represent the values that he ascribes to the four possible outcomes, according to the news value function V_{JONES}. For instance, we know that Jones would be happiest to learn that he will get there on time in a Škoda.[10] Question: what are Jones's V-scores for A and $\neg A$, i.e. for taking the Audi and taking the Škoda?

It is easy to calculate the answer by applying (2.8)(ii), which Cr and V jointly satisfy by definition. Note first that $Cr(\neg L \mid A) = 0.75$ and $Cr(\neg L \mid \neg A) = 0.25$ – this follows from (2.12), (2.13) and the fact that Cr is a probability function. So (2.8)(ii) implies that $V(A) = 0.25 \times -4 + 0.75 \times 2 = 0.5$ and $V(\neg A) = 0.75 \times -2 + 0.25 \times 4 = -0.5$. So Jones must prefer the news that he takes the Audi to the news that he takes the Škoda.

Suppose now that Jones can *choose* between the Audi and the Škoda, the story so far having been neutral on this point. What should he do? It is a natural idea that he should choose to make true whichever proposition would be more welcome news. So he should take the Audi. More generally, we might claim on these grounds that *any* rational person in Jones's position will take the Audi if his V and Cr conform to (2.14)–(2.17).

More generally still: if $O_1, \ldots O_n$ describe an agent's options on any occasion then it is rational to realize an option O_i if and only if it maximizes news value amongst the O_j; that is, if and only if $V(O_i) = \max_j V(O_j)$. Reformulating that in terms of (2.8)(ii) gives:

[10] Of course, these numbers are arbitrary to an extent, since any positive affine transformation of V_{JONES} will represent his news values equally well. What is not arbitrary in this framework is its assignment of interval ratios. For instance, any V such that (Cr_{JONES}, V) satisfies (2.8) will agree with any other over the ratio of the difference in value between being late in the Audi and being late in the Škoda to that between being on time in the Audi and being late in the Škoda.

(2.18) ***Evidential Decision Theory (EDT)***: for any partition $\{S_1, S_2 \ldots S_m\}$ of S, a rationally optimal choice amongst options $O_1, \ldots O_n$ is any that maximizes $V(O_j) = \Sigma_{1 \le i \le m} \, Cr\,(S_i \,|\, O_j) \, V\,(O_j \wedge S_i)$ for $1 \le i \le m$ and $1 \le j \le n$.

In short: maximize news value! This is the doctrine that I defend against its more popular causal rival, to whose motivation and content I'll shortly turn. Before that, two comments on the evidential theory. First: it is *subjective* in the sense of telling us what is rational *by the agent's own lights*, not what she ought to do in some more objective sense, if any exists.

Two features of the formalism reflect this. (i) Most obviously, V itself is subjective in the sense of reflecting the agent's own news preferences, however idiosyncratic or wicked. (ii) The conditional probabilities $Cr\,(S_i \,|\, O)$ are also subjective. They reflect the agent's opinions as to the distribution and strength of the evidential connections between the O_i and the S_j. If Smith thinks that the Audi is no faster than the Škoda then we have $Cr_{\mathrm{SMITH}}\,(L\,|\,A) = Cr_{\mathrm{SMITH}}\,(L\,|\,\neg A)$ but $Cr_{\mathrm{JONES}}\,(L\,|\,A) < Cr_{\mathrm{JONES}}\,(L\,|\,\neg A)$. Only one of them takes Jones's taking the Audi to be evidentially relevant to whether he is late. This *may* – though it need not – reflect error, not just ignorance, on the part of Smith or Jones. If so, it is not the sort of error that EDT is concerned to detect. Rather, EDT will rule that by Smith's lights Jones had better take the Škoda. But by Jones's lights things might be otherwise.

Second: what EDT recommends to the agent is independent of her *causal* beliefs. More precisely, the conditional credences $Cr\,(S_j \,|\, O_i)$, which constitute its doxastic input, depend only on the extent to which the several O_i are deemed by the agent to be evidence for the S_j. Once that is fixed, it makes no further difference to EDT which of these evidential connections are sustained by causal ones. Nor, if a causal relation *does* sustain some such connection, does it matter in which direction it runs. And evidential connections and causal relations will in general not be aligned, since the relation of evidential support is symmetric ($Cr\,(Y\,|\,X) > Cr\,(Y)$ iff $Cr\,(Y\,|\,X) > Cr\,(X)$[11]), whereas causation is generally thought to inherit asymmetry from its temporal orientation. This is hardly surprising. What makes good news good *news* is that it is a *sign* of a good outcome, whether or not it promotes one.

Its being thus indifferent to the agent's causal beliefs is the crucial distinguishing feature of the evidential theory. Some of its supporters take

---

[11] $Cr\,(Y\,|\,X) > Cr\,(Y)$ iff: $Cr\,(X),\, Cr\,(Y) > 0$ and $Cr\,(X \wedge Y) > Cr\,(X)\,Cr\,(Y)$ iff: $Cr\,(X \wedge Y)\,/\,Cr\,(Y) > Cr\,(X)$ iff: $Cr\,(X\,|\,Y) > Cr\,(X)$.

Table 2.2 *Newcomb's problem*

	S_1: being predicts O_1	S_2: being predicts O_2
O_1: you take one box	M	o
O_2: you take both	$M + K$	K

that to count in favour of it.[12] But many philosophers – and most decision theorists – regard it as a fatal flaw. What motivated this conclusion about EDT, and also the generally preferred alternative, was the following sort of example.

2.5　Newcomb's problem

'Suppose a being in whose power to predict your choices you have enormous confidence. (One might tell a story about a being from another planet, with advanced technology and science, whom you know to be friendly, etc.) You know that this being has often correctly predicted your choices in the past (and has never, so far as you know, made an incorrect prediction about your choices), and furthermore you know that this being has often correctly predicted the choices of other people, many of whom are similar to you, in the particular situation to be described below. One might tell a longer story, but all of this leads you to believe that almost certainly this being's prediction about your choice in the situation to be discussed will be correct.'[13]

That situation is as follows. There are two boxes. One box is opaque. The other box is transparent. There is a thousand dollars in it. You have two options. (O_1) Take only the opaque box ('one-boxing'). (O_2) Take the opaque box *and* the transparent box ('two-boxing'). You get to keep whatever is in the box or boxes that you take. Yesterday the being made a prediction about what you now do. If (S_1) it predicted that you take only the opaque box, then it put a million dollars in the opaque box. If (S_2) it predicted that you take both boxes, then it put nothing in the opaque box.

In this problem – called *Newcomb's problem* after its inventor – we may summarize your options, the relevant events and your values for the outcomes in Table 2.2. Here I am writing M for one million and K for one

---

[12] Jeffrey 1983: 157–8; Horwich 1987: 193–4.　　[13] Nozick 1969: 207–8.

thousand, and assuming, as is harmless if unrealistic, that V (you get $\$n$) $= n.$[14] The question is: should you take only the opaque box, or should you take both boxes?

Evidential Decision Theory recommends that you only take the opaque box. Since you know that the predictor is highly accurate, whatever you do is very strong evidence that he has *predicted* that you would do just that. So your $Cr(S_1 \mid O_1)$ and $Cr(S_2 \mid O_2)$ are both high: suppose we set them at $Cr(S_1 \mid O_1) = Cr(S_2 \mid O_2) = 0.9$. Then (2.8) gives:

(2.19) $V(O_1) = 0.9M = 900K$
(2.20) $V(O_2) = 0.9K + 0.1(K + M) = K + 0.1M = 101K$

Since $V(O_1) > V(O_2)$, it follows from (2.18) that it is uniquely rationally optimal to take only the opaque box.

Intuitively this is hardly surprising. Your taking only the opaque box is a very good sign that you are about to become a millionaire. Your taking both boxes is a very good sign that you are *not* about to become a millionaire, if you are not already within a thousand dollars of being one. So obviously any theory that recommends maximizing news value will recommend O_1 over O_2 to anyone who values money.

But it is easy to feel that this is wrong. As Nozick writes:

> The being has already made his prediction, and has already either put the $\$1M$ in the opaque box, or has not. The $\$1M$ is either already sitting in the opaque box, or it is not, and which situation obtains is already fixed and determined. If the being has already put the $\$1M$ in the opaque box, and you take what is in both boxes you get $\$(M + K)$, whereas if you take only what is in the opaque box, you get only $\$1M$. If the being has not put the $\$1M$ in the opaque box, and you take what is in both boxes you get $\$1K$, whereas if you take only what is in the opaque box, you get no money. Therefore, whether the money is there or not, and which it is already fixed and determined, you get $\$1K$ more by taking what is in both boxes rather than taking only what is in the opaque box. So you should take what is in both boxes.[15]

---

[14] The assumption is unrealistic because you realistically have declining marginal news value for money, which is equivalent to risk-aversion for money (Kreps 1988: 72). 'Declining marginal news value for money' means that if G_n is the proposition that you get $\$n$, then $dV(G_n)/dn > 0$ but $d^2V(G_n)/dn^2 < 0$, and this is inconsistent with $\forall n(V(G_n) = n)$. For instance, the sixth or seventh million dollars (say) matters less to most people than the first million, whereas $V(G_n) = n$ implies that each extra million matters as much as the last. The assumption is harmless because it would be possible to do without it. It would just make the exposition more complicated. (For an example of doing without it, see Chapter 4 n. 2.)

[15] Nozick 1969: 208–9, with trivial alterations to fit my terminology.

The force of this objection depends on there being *nothing that you can now do* about what is in the opaque box. To say that there is 'nothing that you can now do about' something is to say that it is *causally* independent of what you do. This follows from its being already 'fixed and determined' what the being predicted that you would do, fixity and determination, whatever *they* are, being (we think) quite general features of the past. Even though your taking only the opaque box is a very good sign that you will be a millionaire, it has no positive *effect* on your wealth. At least, it has none relative to the alternative of taking *both* boxes, which would in any event make you a thousand dollars richer than if you were to take only the opaque box.

So it seems that Evidential Decision Theory gives bad advice in this example. At least, it has seemed that way to many.[16] Its downfall, on this view, is that EDT pays no attention to the *causal* connections between the options and the outcomes. And the proper remedy is a *causal* decision theory, one that advises you to do what causes the outcome that you want.

2.6   K-partitions

Although Newcomb's problem is in one obvious respect an unusual decision problem, one respect in which it is entirely typical is that the agent does not know *what* causal relations hold between the options that he faces and the outcomes that matter to him. If the prediction was that you will take only the opaque box then taking it will cause you to get a million dollars and taking both will cause you to get $\$(1M + 1K)$. If the prediction was that you will take *both* boxes then taking only one will cause you to get nothing; taking both will cause you to get a thousand dollars. But in advance of deciding what to do you don't know what the prediction was. So in advance of deciding what to do, you don't know the effects of your options on your terminal wealth.

It is the same with even the most mundane and straightforward decision situations. For instance, in *Weather Forecast* (Table 1.1), you know that the effect of taking an umbrella is that you will be dry. But you don't know the effect of *not* taking an umbrella. All you know is that *if* it rains then the effect of not taking an umbrella is that you will get wet, and *if* it doesn't rain then the effect is that you will be dry.

But even if you typically don't *know* the effects of your options, you can still take them into account when deciding what to do. The idea is that you have certain degrees of confidence in each of various causal hypotheses, each hypothesis spelling out fully the possible effects of your options. And

---

[16]   In addition to Nozick 1969, see e.g. Gibbard and Harper 1978: 368–72; Lewis 1981b: 309–12; Joyce 1999: 146–54; Weirich 2001: 126–31; Sloman 2005: 89–93.

you can evaluate each option by calculating a weighted average of its value to you on each of these hypotheses, the weights being the degrees of confidence, or credences, that you invest in each hypothesis.

We can divide the set of possible worlds S into exclusive sets of worlds within each of which there is agreement on these fully specifying causal hypotheses. For instance, in Newcomb's problem there are two possible full specifications. These correspond respectively to the state that the predictor has predicted that you will take only the opaque box, and the state that he has predicted that you will take both boxes.

(2.21) K_1: Taking one box will cause you to get \$1M and taking both boxes will cause you to get \$$(1M + 1K)$.

(2.22) K_2: Taking one box will cause you to get nothing and taking both boxes will cause you to get \$1K.

The set $\{K_1, K_2\}$ is a partition of the state space S with respect to your Cr. You are certain that exactly one of K_1 and K_2 is true but you are not certain which.

More generally, let us suppose that in a decision problem we are given a set A of options and set Z of outcomes describing every possible way in which things that the agent cares about could turn out. More formally, Z is *rich* in the sense that for any $X \in Z$ and any Y not in Z, $V(X \wedge Y) = V(X)$. What this means is that once the agent knows which element of Z is true he simply doesn't care what else is true. Thus in Newcomb's problem, if you only care about your terminal wealth, then the set of propositions {You get nothing, You get \$1K, You get \$1M, You get \$$(1M + 1K)$} is a rich partition on S.

Then:

(2.23) For a decision problem with a set O of options and a rich set Z of outcomes, a *K-partition* is a partition on S each element of which fully specifies the causal dependence of each element of Z on each element of O.

Thus consider e.g. *Sink or Swim* (Table 1.4). Here $O = \{O_1$: Learn to swim, O_2: Don't learn to swim\} and (using obvious abbreviations) $Z = \{z_1$: Cost $\wedge$ death, z_2: Cost and survival, z_3: No cost $\wedge$ death, z_4: No cost $\wedge$ survival\}. Assuming that causation is deterministic, it follows that the relevant K-partition K has $4^2 = 16$ elements. Writing $\rightarrow$ for this deterministic connection, it is the set: $\{O_1 \rightarrow z_1 \wedge O_2 \rightarrow z_1, O_1 \rightarrow z_1 \wedge O_2 \rightarrow z_2, \ldots O_1 \rightarrow z_4 \wedge O_2 \rightarrow z_3, O_1 \rightarrow z_4 \wedge O_2 \rightarrow z_4\}$ – or rather, it is whatever subset of that set you get if you subtract all of the causal hypotheses in which Alice has zero credence. For instance, she may have zero credence on any hypothesis according to which learning to swim costs nothing. In that case, no proposition that entails either $O_1 \rightarrow z_3$ or $O_1 \rightarrow z_4$ belongs to the K-partition.

Of course it may be that the universe is not fully deterministic. It may be, for instance, that neither its own past, nor even the entire history of the universe up to this point, causally determines whether this radium atom decays in the next five minutes. So there might be some not-certainly-false causal hypothesis on which some or all of the options available to the agent do not fully causally determine whether this or that outcome occurs, even against the fixed background of the entire history of the world up to the time of acting. In that case the elements of the K-partition do not only include propositions asserting full causal determination of outcomes by options. They also include propositions stating that this or that option gives this or that option this or that *chance*.

Let me illustrate this point before saying more about it in general terms. This button is connected to a bomb via a variable resistor and a Geiger counter: current only flows through the circuit if (a) you press the button and (b) the Geiger counter registers radioactivity above a certain threshold. A radioactive source is placed nearby: this is either a very unstable or a relatively stable isotope of some radioactive element. Your options are to press the button (A_1) and not to press it (A_2); the only thing that concerns you is whether or not (E) the explosion takes place. Assuming that radioactive decay is an indeterministic process, it may be that no element of the K-partition entails that pressing the button will causally determine that an explosion occurs. Rather, the two hypotheses will be something like this:

(2.24) K_1: Pressing the button gives the explosion a *low* chance and not pressing it causes the explosion not to occur.

(2.25) K_2: Pressing the button gives the explosion a *high* chance and not pressing it causes the explosion not to occur.

Here K_1 corresponds to the possibility that the radioactive material consists of the relatively stable isotope and K_2 corresponds to the possibility that it consists of the very unstable isotope. Of course this is only a first approximation, because, as stated, K_1 and K_2 make only non-numerical estimates of the chances of an explosion, given that you press the button.

So as to make things more numerically definite I'll introduce the notions of unconditional and conditional **chance**. Chance is a measure of *objective* probability: it associates with each time a probability function Ch_t whose value, for a particular proposition, reflects the causal strength with which the state of the world, up to that time, promotes its *coming* true. So for instance, if H_{t-1} is the proposition that a fair coin landed heads last time it was tossed, and $t =$ now, then $Ch_t (H_{t-1}) = 1$ or 0 depending on whether F_{t-1} is true or false. But if H_{t+1} is the proposition that it will land heads on its *next* toss then $Ch_t (H_{t+1}) = 0.5$ whether H_{t+1} is true or false.

The chance function at any time is a wholly objective feature of reality and quite independent of anyone's beliefs. So, for instance, if my credences are equally and exhaustively divided between the proposition that this coin has heads on both sides and the proposition that it has tails on both sides, then my credence that it will land heads on the next toss might be 0.5. But if the coin *does* have either two heads or two tails then the present *chance* of heads on the next toss cannot be 0.5. It must be 0 or 1.

Similarly, the conditional chance $Ch_t (Q|P)$, t being just before the time that P describes, measures the objective extent to which P (or the event that it describes) causally promotes Q in just the way that $Cr (Q|P)$ measures the extent to which P is subjective evidence for Q: P causally promotes Q if and only if $Ch (Q|P) > Ch (Q|\neg P)$. And if Q is causally *independent* of P then $Ch (Q|P) = Ch (Q)$.

The key connection between objective chance and subjective credence is the following:

(2.26) **Principal Principle**: Given any time t and evidence E admissible at t, if Cr is a reasonable subjective credence function and P and Q are propositions then for any $x \in [0, 1]$:
  (i) $Cr (P|E \wedge Ch_t (P) = x) = x$
  (ii) $Cr (Q|P \wedge E \wedge Ch_t (Q|P) = x) = x$[17]

In the applications that matter here, t is the present time and P describes options, i.e. acts, that are presently available to the agent whose beliefs Cr describes. So I will write Ch for Ch_t if $t =$ now. Saying that the evidence E is 'admissible at t' means that E doesn't concern the post-t evolution of whatever chance process leads up to what P (in (i)) or Q (in (ii)) describe. For instance, the agent is not allowed to use crystal balls etc. to learn in advance the outcome of the next toss. What all of this implies is e.g. that given a fair coin you should have a credence of 0.5 that it lands heads if tossed, assuming that no fortune-teller has told you how it will in fact land. And if you are 50/50, concerning this other coin, between its having two heads and its having two tails, you should again have credence 0.5 that it lands heads on its next toss.

Returning to the K-partition: in indeterministic cases its elements are full specifications of the conditional chances of outcomes on options, as these are just before the moment of action. The agent divides her credence over

[17] (2.26)(i): Lewis 1980: 86–7. (2.26)(ii): Joyce 1999: 166.

these in some way, these credences corresponding to the weights assigned to the outcomes in a manner that I will shortly make explicit. For instance, in the case involving the button and the radioactive source, there are two possibilities. If the source is unstable, the chance of (E) an explosion given (A_1) that you press the button is high. If the source is stable, the chance of an explosion given that you press the button is low. In either case (A_2) your not pressing the button will cause no explosion at all. So if t is the time of decision, then the elements of the K-partition might (for instance) be:

(2.27) $Ch_t\,(E\,|\,A_1) = 0.8 \wedge Ch_t\,(E\,|\,A_2) = 0$
(2.28) $Ch_t\,(E\,|\,A_1) = 0.2 \wedge Ch_t\,(E\,|\,A_2) = 0$

In every case so far the K-partition is finite. But once we grant that the elements of the K-partitions are specifications of conditional chances we cannot expect this to happen in general; a K-partition might even be uncountable. That will happen if, for instance, one is considering whether to toss a coin of completely unknown bias and cares only whether it lands heads: here the K-partition $\{Ch_t\,(\text{Heads}\,|\,\text{Toss}) = n\,|\,0 \leq n \leq 1\}$ is the size of the continuum. And similarly one's credence need not be discretely apportioned amongst the members of the K-partition: in the present case, for instance, it may be that what represents your beliefs about the result of the toss is a probability density function.

But most actual applications of Causal Decision Theory that will matter for present purposes all involve simple problems with finite K-partitions. I turn now to the content of the theory itself.

2.7   Causal Decision Theory

We are finally in a position to give a formal statement of the causal theory.

(2.29)   *Causal Decision Theory (CDT):* For a decision problem with a K-partition K and a set of options O, an option $O \in O$ is rational if and only if it maximizes the *utility function* U over O: $U\,(O)$ $=_{\text{def.}} \Sigma_{K \in K}\, Cr\,(K)\, V\,(O \wedge K)$

In short, Causal Decision Theory advises agents to maximize *utility*, where the utility of an option is the weighted average of its news values on each full causal hypothesis, the weights being the credence that the agent attaches to each such hypothesis.[18]

---

[18] In the case that K is not countable it is necessary to replace the discrete summation with an appropriate integral: for an application see Chapter 5 n. 13 below.

Newcomb's problem illustrates both the content of this definition and the difference between utility and news value. Supposing that you care only about terminal wealth, we can write $K = \{K_1, K_2\}$, where the elements of the K-partition are as at (2.21) and (2.22), these corresponding to the hypothesis that the predictor has predicted that you would take one box and that he has predicted that you would take both boxes. The options in Table 2.2 therefore get the following U-scores:

(2.30) $\quad U(O_1) = Cr(K_1) V(O_1 \wedge K_1) + Cr(K_2) V(O_1 \wedge K_2) = Cr(K_1)$
$\quad\quad\quad V(\$1M) + Cr(K_2) V(\$0)$

(2.31) $\quad U(O_2) = Cr(K_1) V(O_2 \wedge K_1) + Cr(K_2) V(O_2 \wedge K_2) = Cr(K_1)$
$\quad\quad\quad V(\$(M+K)) + Cr(K_2) V(\$1K)$

Assuming only that you have increasing news value for dollar pay-offs, it follows from (2.30) and (2.31) that $U(O_2) > U(O_1)$, quite regardless of the actual values of your $Cr(K_1)$ and $Cr(K_2)$, i.e. quite regardless of what you think the predictor has put in the opaque box. Propositions (2.30) and (2.31) thus represent the formal counterpart of the dominance argument: whatever amount is already in the opaque box, you are better off taking both boxes. So you should two-box.

Contrast (2.30) and (2.31) with (2.19) and (2.20), according to which Evidential Decision Theory recommends one-boxing. The difference that accounts for this disagreement is that CDT weights outcomes of options by the *un*conditional probabilities of the full causal hypotheses that lead to them, whereas for EDT the correct weights reflect the evidential bearing of each option upon the causal hypotheses themselves. Thus instead of $U(O)$ as defined at (2.29), EDT concerns itself with:

(2.32) $\quad V(O) = \Sigma_{K \in K} Cr(K \mid O) V(O \wedge K)$

The difference between EDT and CDT is therefore essentially the difference between the term $Cr(K)$ as it appears in (2.29) and the term $Cr(K \mid O)$ as it appears in (2.32).

And that is what you would expect. Each element of the K-partition is a *full* causal story about the effects of acting in this or that way, in so far as the effects matter. Any difference between $Cr(K)$ and $Cr(K \mid O)$ must therefore reflect the residual evidential bearing of O upon K, and hence upon the outcomes, via channels that do not reflect any causal influence of O upon these outcomes. So it is hardly surprising that Evidential Decision Theory, and only Evidential Decision Theory, should care about this residue.[19]

---

[19] Lewis 1981b: 314.

2.8 Matters arising

I should briefly mention five matters arising from this exposition of EDT and CDT. First, I have said nothing about what counts as an option; that is, which propositions in the field of $\succ$ are 'available' to the agent. Do they describe events that depend wholly on the agent's will, for instance? Or do they describe ordinary bodily movements, which depend for their occurrence not only on the agent's brain-state but also on the co-operation of his extra-cranial physiology? Or do they describe the apparently extra-*bodily* events to which our ordinary talk of actions sometimes seems exclusively to refer, as when e.g. 'The queen killed the king' entails the death of the king but not any bodily motion of the queen, such as the movement of her hand as she pours poison in his ear?

It makes a difference. Given suitable background beliefs and desires, CDT and EDT might both prefer killing the king to not killing him, and yet also prefer *not* pouring poison in his ear to doing so.[20] So the question how to formulate the decision problem, in particular what to consider an available option in the first place, remains open and pertinent.

But we needn't pursue it very far. Here I shall take options to be propositions describing bodily movements that are normally available to the agent. This *includes* such propositions as 'The queen kills the king' if, following Davidson, we take all descriptions of options to be descriptions (perhaps indirect and perhaps inter alia) of bodily movements.[21] In any case it makes no difference to the cases I'll consider: there, all available options are expressed in terms that make no reference to events whose connection with the relevant bodily movement or act of will is in question. For instance, if it is an *option* for Jones that he rents an Audi, then it is in a decision situation where there is no question that Audis are available, that he can afford to rent one, etc.[22]

---

[20] Thus suppose that the chance of the king's death given poisoning is 0.1 but that the chance of detection is then 0.5. Detection is irrelevant if the king dies but disastrous if he doesn't. In that case it might be a bad idea to poison the king but a good idea to kill him. EDT and CDT may both give different advice depending on which of these is taken as the available act. The issue arises in connection with an alleged version of Newcomb's problem: see Chapter 4 n. 11 below.

[21] Davidson 1971.

[22] Joyce (1999: 57–67); and Hedden (2012) identify options with purely mental acts of will or decisions. Generally this makes no difference to what follows, although it *may* be a difficulty for this view that it has the following consequence: if determinism is true then one is free to do something that would either be or cause some violation of a law of nature. See Beebee 2003, Beebee and Mele 2002 and Ahmed 2013: 290 n. 2. This is perhaps the appropriate place to mention the *ratificationist* version of EDT that Jeffrey adopted in response to the Newcomb case (Jeffrey 1983: 15 ff.). Jeffrey distinguishes between *choosing* an option and *performing* it: it is crucial to this that one might choose an option that one does not in the end perform, for instance because death or a nonfatal cerebral haemorrhage

Second, I haven't said anything about causation itself, other than the assertion that there is a chance function of which the distribution conditional upon a proposition reflects the causal effects of what the proposition describes. One objection to this is that a pre-empted putative cause may be causally irrelevant to the effect of the pre-emptor, despite raising the chance of that effect at the time of its own occurrence. Thus if two stones A and B are thrown simultaneously at a vase, each of which would, if thrown on its own, have a 0.5 chance of smashing the vase, then the throwing of B raises the chance that the vase will smash even if it was A that actually smashed the vase on this occasion.[23] But being a subjective theory, Causal Decision Theory concerns not the causal or chance-involving facts themselves but rather the agent's beliefs about them. An agent that lacked inadmissible evidence at the time of choosing whether to throw a stone would not then *believe* that P (that she throws the stone) is causally irrelevant to Q (that the vase smashes) unless she also thought that P does nothing to raise the chance of Q. So it is legitimate to take chance-raising to imply causal relevance in the context of a *subjective* decision theory that assesses options against the background of the agent's subjective credences, whether or not these reflect the actual causal relations.[24]

Still, it remains the case that the agent's opinions of the conditional chance distribution, and so his division of credence over any K-partition,

intervenes (ibid.: 18). An option O is then said to be ratifiable if its *performance* maximizes news value, compared to the news value of the performances of the other options, on the hypothesis that O is chosen, this quantity being, for any given option O', $V_O(O') =_{\text{def.}} \Sigma_{1 \le i \le m} Cr(S_i | O' \wedge dO) V(O' \wedge dO \wedge S_i)$, where $\{S_1, \dots S_m\}$ is any partition on S and $dO =_{\text{def.}}$ the proposition that one has *chosen* O. And Jeffrey adds to EDT the maxim that one should only choose the V-maximal performance from amongst the *ratifiable* options. This form of evidentialism no longer recommends one-boxing in at least some versions of the Newcomb problem. Suppose that learning that you have decided to one-box is strong evidence, evidence that the actual performance does nothing to augment, that this is what the predictor predicted that you would do. Then in the terms of Table 2.2, we have $V_{O_1}(O_1) \approx V(M) < V_{O_1}(O_2) \approx V(M + K)$; but $V_{O_2}(O_2) \approx V(K) > V_{O_2}(O_1) \approx V(0)$: so two-boxing, and only two-boxing, is ratifiable. But if we take the options to be the choosings themselves then this emendation to EDT is unavailable, since it is hardly clear what it is to *choose to choose* to do something, unless that just means to choose it, in which case the cross-values $V_O(O')$ makes no sense when $O \ne O'$. In any case and as Jeffrey admits, there are versions of Newcomb's problem in which the performance of a bodily movement does have merely symptomatic significance for the state of interest independently of one's choosing it (Jeffrey 1983: 20; see also Chapter 4 n. 11, Chapter 4 n. 15 and section 4.6 below). For further discussion of other forms of ratificationism, see the discussion of the 'Full Information' and 'Piaf' principles at sections 3.1 and 3.2 below.

23 Hitchcock 2004: 410.
24 A similar point applies to D. Rosen's counterexample to causation as chance-raising: a golfer slices a ball, which bounces off a tree for a hole in one. The slice caused the hole in one but *lowered* its chance of happening (Rosen 1978: 607–8). But *ex ante* the golfer would not be in any position to know this: as far as he could tell, it would seem likely both that slicing would lower the chance of a hole in one *and* that slicing would causally retard this outcome. See further Chapter 8 n. 2 below.

will at least sometimes depend on his prior beliefs about causal dependence and independence. In most of the cases studied here these beliefs are either stipulated or so obvious that we may take them for granted. For instance, in Newcomb's problem, we simply take for granted that the prediction is causally independent of what you now do.[25] The main exceptions appear in Chapter 6, where the examples have a controversial causal structure on which plausible analyses of causation do not agree. I will discuss the matter separately there (see section 6.5.1).

Third, the present discussion of CDT has ignored the fact that there are *various* formulations of Causal Decision Theory. The present formulation is essentially Lewis's. Skyrms endorses something similar, and Joyce regards it as an acceptable formulation of CDT.[26] But some writers formulate that theory in counterfactual rather than in chance-raising terms.[27] For them, what should concern the rational agent is not the conditional chance of this or that outcome given this or that act, but rather whether this or that outcome *would* obtain if the agent *were* to perform this or that act.

But none of these differences make a difference: in all of the cases that follow, they and the present formulation, and also I think *anything* worth calling a *causal* decision theory, are in agreement with one another as to what the agent should *do*. If any one of them is true then causalism is true; if any of them gets the following examples wrong then they all do. As far as this discussion is concerned, any one of them could stand for their disjunction.

Fourth, the requirement of richness means that the implementation of CDT is altogether more demanding than the implementation of EDT. To say that a partition is rich is to say that once the agent knows which element of the partition is actual, he is indifferent to any further news. So every element of a rich partition must specify everything that matters to the agent. Typically this will include a great deal besides the net financial upshot of the transaction: for instance, all his future welfare and decisions, as well as those of anyone who matters to him, also the outcomes of sporting fixtures and political processes to which he is not indifferent. So in any real-life case it would be very hard to specify a rich partition.

We can't simply drop the requirement that Z be rich from the definition of K-partition at (2.23), thus allowing *any* partition to play the role of Z in that definition. To see how badly wrong that would be, suppose

[25] For an objection to that, see Price 2012: 507–11. I discuss the 'agency' theory that underlies this objection at section 5.3.1 below.

[26] Skyrms 1980: 133; Lewis 1981b: 313; Joyce 1999: 161. For an argument that Lewis's theory reduces to the version outlined here, see ibid.: 173 n. 45. Joyce himself prefers a different, counterfactual formulation of Causal Decision Theory (Joyce 1999: 172), but his reasons for this have got nothing to do with the actual recommendations of any chance-raising formulation.

[27] Gibbard and Harper 1978; Sobel 1986: 254–8.

that in Newcomb's problem we simply chose $Z = \{O_1, O_2\}$. Then the corresponding K-partition has just one element: $K = \{Ch\,(O_1 \mid O_1) = 1 \wedge Ch\,(O_2 \mid O_2) = 1\}$, because the chances of the elements of *this* Z can causally depend on what you do in only that way. It follows from (2.29) that the U-scores and the V-scores of the options coincide, and CDT recommends one-boxing. Just dropping richness subverts the whole point of the causal theory.

In fact we can make do with a weaker but more complicated condition than richness. Let Z be a genuinely rich partition on S. Then suppose that we have a relatively gross partition Z^* on S such that (i) $z \subseteq z^*$ if z and z^* are not disjoint and $z \in Z$ and $z^* \in Z^*$ (so that every cell of Z^* is a union of cells of Z); and (ii) no option has any evidential bearing on its own causal influence on Z given its causal influence on Z^*. In that case Z^* will serve as a suitable set of outcomes for the purposes of applying Causal Decision Theory.[28]

For instance, if in Newcomb's problem we suppose that (i) how much money you make from this exercise necessarily makes a difference to your total welfare; (ii) which box you take makes no difference to your total welfare except via its effect on your profit from this exercise – if we suppose these not at all implausible things, then we may legitimately apply CDT to the Newcomb problem without having to imagine that your monetary reward from this situation exhausts everything that you care about.

But since we *can* do without richness, there is no harm in assuming it; and for the sake of simplicity that is what I will do. That is, in all of the problems to which I apply CDT, I shall do so on the implausible assumption that the partition whose elements specify the terminal wealth of the agent is indeed rich. Implausible it is, but harmless too, since we know that CDT would make the same recommendations on the more complex but also more plausible assumptions that I just outlined.[29]

---

[28] Let K be the K-partition whose elements specify the causal dependencies of elements of Z on your options $O \in O$. Let K^* be the K-partition whose elements specify the causal dependencies of elements of Z^* on your options. Then the assumptions are:

(i) $z \subseteq z^*$ if z and z^* are not disjoint and $z \in Z$ and $z^* \in Z^*$

(ii) $Cr\,(K \mid O \wedge K^*) = Cr\,(K \mid K^*)$ for any $K \in K$, $K^* \in K^*$, $O \in O$; so:

(iii) $\Sigma_{K^* \in K^*}\, Cr\,(K^*)\, V\,(O \wedge K^*) = \Sigma_{K^* \in K^*}\, Cr\,(K^*)\, \Sigma_{K \in K}\, Cr\,(K \mid O \wedge K^*)\, V\,(K \wedge O \wedge K^*)$ by (2.8)(ii)

(iv) $\Sigma_{K^* \in K^*}\, Cr\,(K^*)\, V\,(O \wedge K^*) = \Sigma_{K^* \in K^*}\, Cr\,(K^*)\, \Sigma_{K \in K}\, Cr\,(K \mid O \wedge K^*)\, V\,(O \wedge K)$ by (i), (iii)

(v) $\Sigma_{K^* \in K^*}\, Cr\,(K^*)\, V\,(O \wedge K^*) = \Sigma_{K^* \in K^*}\, Cr\,(K^*)\, \Sigma_{K \in K}\, Cr\,(K \mid K^*)\, V\,(O \wedge K)$ by (ii), (iv)

(vi) $\Sigma_{K^* \in K^*}\, Cr\,(K^*)\, V\,(O \wedge K^*) = \Sigma_{K \in K}\, Cr\,(K)\, V\,(O \wedge K)$ by (v)

(vii) $\Sigma_{K^* \in K^*}\, Cr\,(K^*)\, V\,(O \wedge K^*) = U\,(O)$ by (vi), (2.29)

[29] Joyce (1999: 176–80) approaches this problem in a different way. His formulation of CDT settles the utility of an option relative to a given partition in a way that is partition-invariant in the sense

Finally, let me comment on the relations between EDT and CDT on the one hand, and Savage's theory on the other. Comparison of (1.22) and (2.18) reveals that whereas SEU weights the probability of a state–act pair by the *un*conditional probability of the state, EDT weights it by the probability of the state *conditional* on the act. This difference vanishes if the states $S_1, \ldots S_n$ are supposed to be *evidentially independent* of the act, for in that case $Cr(S_i \mid O) = Cr(S_i)$, and the formula for news value reduces to:

$$(2.33) \quad V(O) = \Sigma_{1 \le i \le n} Cr(S_i) V(O \wedge S_i)$$

And *V*-maximization coincides with SEU-maximization. In particular, EDT endorses Savage's dominance principle (1.17) provided that the states are evidentially independent of the acts. At section 7.4.1 we shall see a case of evidentially independent but not causally independent states to which only this evidential version of the dominance principle applies.

Similarly, CDT coincides with SEU-maximization if the states in the latter are taken to be *causally* independent of the agent's acts. More precisely, if O is your set of options and $P = \{P_1, \ldots P_n\}$ is any partition over S such that (i) $O \otimes P =_{\text{def.}} \{O \cap P \mid O \in O, P \in P\}$ is rich and (ii) P is causally independent of O, then:

$$(2.34) \quad U(O) = \Sigma_{1 \le i \le n} Cr(P_i) V(O \wedge P_i)^{[30]}$$

of section 2.2 above, so that it makes no difference whether one calculates this utility relative to a very fine partition or a relatively coarse one. But Joyce's general definition (ibid.: 178) still relies on a specification of the utilities of individual possible worlds, or at least of elements of some genuinely rich partition. Without these it is impossible to apply the theory. So it is not clear that Joyce's approach really solves the problem.

[30] Proof: suppose given some partition Q such that $O \otimes Q$ is rich (Q possibly distinct from P) and a K-partition K such that each $K \in K$ specifies the chances of each $Q \in Q$ conditional on each option O. Let K^* be a partition whose elements each fully specify the causal dependencies of elements of P on elements of O. Then for arbitrary $O \in O$ we have:

(i) $U(O) = \Sigma_{K \in K} Cr(K) V(O \wedge K)$ by (2.29)

(ii) $U(O) = \Sigma_{K \in K} \Sigma_{K^* \in K^*} Cr(K \wedge K^*) V(O \wedge K)$ by (i)

(iii) $U(O) = \Sigma_{K \in K} \Sigma_{K^* \in K^* \text{ s.t. } Cr(K \wedge K^*) > 0} Cr(K \wedge K^*) V(O \wedge K)$ by (ii)

(iv) $V(O \wedge K) = \Sigma_{Q \in Q} Cr(Q \mid O \wedge K) V(Q \wedge O \wedge K)$ by (2.8)(ii)

(v)   For any $K \in K, K^* \in K^*$ s.t. $Cr(K \wedge K^*) > 0$, $V(O \wedge K) = \Sigma_{Q \in Q} Cr(Q \mid O \wedge K \wedge K^*) V(Q \wedge O \wedge K)$ by (iv) and (2.26)(ii)

(vi)   For any $K \in K, K^* \in K^*$ s.t. $Cr(K \wedge K^*) > 0$, $V(O \wedge K) = \Sigma_{Q \in Q} Cr(Q \mid O \wedge K \wedge K^*) V(Q \wedge O \wedge K \wedge K^*)$ by (v) and the richness of $O \otimes Q$

(vii)   For any $K \in K, K^* \in K^*$ s.t. $Cr(K \wedge K^*) > 0$, $V(O \wedge K) = V(O \wedge K \wedge K^*)$ by (vi) and (2.8)(ii)

(viii) $U(O) = \Sigma_{K \in K} \Sigma_{K^* \in K^* \text{ s.t. } Cr(K \wedge K^*) > 0} Cr(K \wedge K^*) V(O \wedge K \wedge K^*)$ by (iii), (vii)

(ix)   For any $K \in K, K^* \in K^*$ s.t. $Cr(K \wedge K^*) > 0$, $V(O \wedge K \wedge K^*) = \Sigma_{P \in P} Cr(P \mid O \wedge K \wedge K^*) V(O \wedge P \wedge K \wedge K^*)$ by (2.8)(ii)

(x)   For any $K \in K, K^* \in K^*$ s.t. $Cr(K \wedge K^*) > 0$, $V(O \wedge K \wedge K^*) = \Sigma_{P \in P} Cr(P \mid K \wedge K^*) V(O \wedge P \wedge K \wedge K^*)$ by (ix), (2.26)(ii) and the fact that P is causally independent of O

(xi)   For any $K \in K, K^* \in K^*$ s.t. $Cr(K \wedge K^*) > 0$, $V(O \wedge K \wedge K^*) = \Sigma_{P \in P} Cr(P \mid K \wedge K^*) V(O \wedge P)$ by (x) and the fact that $O \otimes P$ is rich

Formula (2.34) will be very useful: in many cases it applies more simply than the official definition of utility from (2.29). The exceptions, which justify the more cumbersome approach in terms of K-partitions and conditional chances, include indeterministic cases like the quantum-mechanical situations that I cover in Chapter 6.

But the present point is that if in Savage's formalism we restrict the set of states to partitions that are causally independent of the options, U-maximization and SEU-maximization coincide over options. In particular the principle of dominance holds for states that are causally independent of the agent's options. Newcomb's problem itself is an example of this, and we shall see many more in the sequel.

In fact not only EDT and CDT, but also many other decision rules, are describable as alternative interpretations of elements in the definition of Savage's SEU.[31] This includes Maximin (optimize the worst possible outcome),[32] Maximin Expected Utility (maximize the worst expected utility on some family of probability distributions)[33] and Minimax Regret (minimize how much better you could have done in your actual circumstances).[34] This flexibility vindicates the continuing centrality of Savage's construction to modern expositions of decision theory.

My reason for not discussing those other rules here is that nothing in the issues amongst them, or between any of them and EDT or CDT, illuminates the philosophical question whether *causal* knowledge is requisite for practical deliberation. The best decision-theoretic crystallization of *that* issue is the dispute between EDT and CDT: if causal knowledge has a special place in practical deliberation then it has the place that CDT gives it.

In fact this claim has been contested: there exist philosophical arguments that CDT misdirects the normative force of specifically causal knowledge because that force, whilst genuine, either (a) has some other proper operation in subjective decision theory or (b) belongs outside *subjective* decision theory altogether. I'll turn now to three arguments of this sort that either are important or have been influential.

(xii) $U(O) = \Sigma_{K \in K} \Sigma_{K^* \in K^* \text{ s.t. } Cr(K \wedge K^*) > 0} Cr(K \wedge K^*) \Sigma_{P \in P} Cr(P | K \wedge K^*) V(O \wedge P)$
by (viii), (xi)

(xiii) $U(O) = \Sigma_{P \in P} \Sigma_{K \in K} \Sigma_{K^* \in K^* \text{ s.t. } Cr(K \wedge K^*) > 0} Cr(K \wedge K^*) Cr(P | K \wedge K^*) V(O \wedge P)$
by (xii)

(xiv) $U(O) = \Sigma_{P \in P} Cr(P) V(O \wedge P)$ by (xiii)

---

[31] Chu and Halpern 2004.   [32] Resnik 1987: 28–32.
[33] Gilboa and Schmeidler 1989.   [34] Resnik 1987: 26–7.

CHAPTER 3

Causalist objections to CDT

The main argument is now as follows. (1) If the agent's *causal* beliefs are relevant to assessing his options then they play the role that Causal Decision Theory says they do. (2) Causal belief does not play this role. Therefore (3) causal knowledge is unnecessary for practical deliberation. Chapters 4–7 argue for the second premise. This chapter argues for the first.

It counts in favour of (1) that if, in the cases reviewed, the agent's causal belief is relevant in any way at all, then it is relevant in the way that CDT says it is. For instance, it is clear that in *Jones and the Auto Rental* (Table 2.1) the only cause–effect relations that need concern Jones are those connecting his choice of rental to (a) his arrival and (b) his wealth. And it is plausible that if they do concern Jones then they should concern him in the way that CDT says they do. He should weight the value he attaches to taking the Audi, and its causing him to get to Swindon on time, by his credence that taking the Audi does cause him to get there on time.

But there are other cases where CDT allegedly goes wrong, not because it gives *some* weight or role to the agent's causal beliefs, but because it gives them the *wrong* weight or role. The details, and my responses, are at sections 3.1 and 3.2 below.

Section 3.3 discusses a more fundamental objection, not only to (1) but also to my whole approach. I am here presupposing that given any beliefs and values regarding some situation, we can ask whether it is rational to realize this or that option in the light of those beliefs and values. Mellor rejects this. For him, the rationality of an option depends only on the objective facts (including the causal facts) of the agent's situation. If he is right then it is pointless arguing for the normative superiority of one subjective decision theory over another, because subjective decision theory is *never* normative.

Table 3.1 *Psycho Button A*

	P: psychopath	¬*P*: normal
A: you push button A	– 90	10
Z: you do nothing	0	0

3.1 Egan–Gibbard

Gibbard and Egan have drawn attention to a family of cases in which, according to Egan, CDT gives plainly counterintuitive advice.[1] The following example is particularly compelling.

> *Psycho Button A*: You can push this button (button A), and you can do nothing. Pushing button A will cause all psychopaths to die. Doing nothing causes nothing. You want to live in a world without psychopaths. You want more not to die yourself. If you do push button A, it is 99 to 1 that you are a psychopath. This is not because doing so makes you one (it doesn't) but because only a psychopath would do such a thing. Your options, a relevant state-partition over *S*, and your values for the outcomes are therefore as in Table 3.1. Finally, you are 95 per cent confident that you are not a psychopath. What do you do?

CDT advises that you should press the button. Since you are 95 per cent confident that you are not a psycho, you are 95 per cent confident that the *effect* of pushing button A would be to put you in the top right-hand cell of Table 3.1; given the news values of doing so in that case, that makes it right to do so. A simple calculation confirms this: since $\{P, \neg P\}$ is causally independent of $\{A, Z\}$, (2.34) gives:

$$(3.1) \quad U(A) = 0.05\,(-90) + 0.95\,(10) = 5$$
$$(3.2) \quad U(Z) = 0$$

So *A* has greater utility than *Z*, and CDT advises *A* over *Z*.

Egan rejects this on the basis that anyone who pushes button A is almost certainly a psychopath to whom doing so is fatal.[2] His view, for which he claims intuitive support, is that:

$$(3.3) \quad \text{*Z* is rationally superior to *A* in a straight choice between them.}$$

Similarly in all cases with this structure, CDT recommends the option that is (a) likely to cause the best outcome but (b) itself a sign that it will

---

[1] Gibbard 1992; Egan 2007.　　[2] Egan 2007: 97.

Table 3.2 *Psycho Button B*

	P: psychopath	¬*P*: normal
B: you push button B	− 90 − Δ	10 − Δ
Z: you do nothing	0	0

cause the worst.[3] CDT responds only to (a). But on Egan's view, decision theory should also be sensitive to (b). Since (a) and (b) are both essentially causal claims, the objection is not that CDT *has* specifically causal concerns but that it is wrongly indifferent to a certain type of causal information. In particular:

(3.4)   Decision theory should care not only about the causal *effects* of a choice but also about its *evidential* bearing on *what those effects are*.[4]

If (3.4) is correct then my conditional premise (1) is false. In that case *neither* EDT *nor* CDT is right. CDT is false because it gets things wrong in *Psycho Button A*. EDT gets that case right by Egan's lights: it is easily checked that $Z \succ A$. But it too violates (3.4), not because it misdirects the agent's causal beliefs but because it ignores them altogether. This comes out in the consequence of (3.4), to which Egan explicitly assents,[5] that EDT goes wrong in the Newcomb problem.

But (3.4) is untenable. The reason is that (3.3) is not *co*-tenable with two-boxing in the Newcomb problem: but as I now argue, (3.4) enforces both of these things.

To this end, consider:

> *Psycho Button B*: You can push this button (button B), and you can do nothing. Pushing button B will cause all psychopaths to die. Doing nothing causes nothing. You want to live in a world without psychopaths. You want more not to die yourself. But the pushing of button B is *not* symptomatic of being a psychopath, nor does it make you one. Still, it does involve a small payment whose disutility to you is Δ, where $0 < Δ < 5$. So your situation is as in Table 3.2. As before, you are 95 per cent confident that you are not a psychopath. What do you do?.

With your credences as before we have: $U(B) = 5 − Δ$ and $U(Z) = 0$. Since $5 > Δ$, CDT therefore recommends that you push button B. And the

---

[3] For instance: you have an opportunity to shoot your rival, but pulling the trigger is evidence that you are a bad shot: Egan 2007: 97. Or: smoking, which you would enjoy, causes a disease iff you have a gene that makes you likely to smoke: Gibbard 1992: 218.
[4] Egan 2007: 101–2.   [5] Egan 2007: 93–6.

Table 3.3 *A or B?*

	P: psychopath	¬*P*: normal
A: you push button A	– 90	10
B: you push button B	– 90 – Δ	10 – Δ

advocate of (3.4) should agree. After all, in this case your choice has *neither* any causal relevance on whether you are a psychopath *nor* any evidential relevance to its own effects. So Egan should grant:

(3.5)   *B* is rationally superior to *Z* in a straight choice between them.

Next, consider:

> *A or B?* This time you get to choose between pushing button A and pushing button B – but you must push one of them. Pushing either button will kill all psychos. If you *are* a psycho then you are very likely to push A and not B; conversely *non*-psychopaths tend to push B. More precisely: if you push A then it is 99 to 1 that you are a psycho. If you push B then it is practically certain that you're not. Pushing B incurs the same cost as in *Psycho Button B*. So your situation is as in Table 3.3. As before you are 95 per cent confident that you're *not* a psycho. What should you do?

The previous calculations apply to this case and imply that $U(A) = 5$ and $U(B) = 5 - \Delta$. Since $\Delta > 0$, CDT prefers *A* to *B*. That is not surprising: nothing that you can do in *A or B affects* whether you are a psycho. But if you *are* a psycho then you are better off (by Δ) pushing A. And if you are *not* a psycho then you are *still* better off pushing A. So of course CDT will advise you to push A.

And Egan should agree, because *A or B* is a version of Newcomb's problem. That is: one option dominates the other (is better in all events) relative to a partition {*P*, ¬*P*} of *S* that is causally independent of your choice. So anyone who advocates two-boxing in Newcomb's problem must advocate pushing button A in *A or B*.[6] So Egan and other supporters of (3.4) must sign off on:

---

[6] It might seem a significant difference between *A or B* and Newcomb's problem that in *A or B*, option A is evidentially relevant to what its own effects are, whereas neither option has that feature in Newcomb's problem. So it may seem that anyone who grants (3.4) should reject (3.6). But that is not so: in *A or B* it is *not* true that pushing button A is evidence that doing so will make any difference to your survival. For instance, if you are a psychopath who pushes button A then you would have died whichever option you had taken in *A or B*. Similarly in Newcomb's problem, taking only the opaque box is evidence that you will be a millionaire, but it is not in the Bayesian sense *evidence* that it would make any difference whether you took only that box or also the transparent box.

Table 3.4 *ABZ*

	P: psychopath	¬*P*: normal
A: you push button A	− 90	10
B: you push button B	− 90 − Δ	10 − Δ
Z: you do nothing	0	0

(3.6)   *A* is rationally superior to *B* in a straight choice between them.

So unlike CDT (which rejects (3.3)), or EDT (which rejects (3.5) and (3.6)), the advocate of (3.4) is committed to *all* of (3.3), (3.5) and (3.6). That is: he must say that in straight choices between them, *Z* is superior to *A*, *A* is superior to *B*, and *B* is superior to *Z*. This may lead to trouble by itself. Certainly it will if you think that pair-wise preference must be transitive.[7] In any case it makes trouble in:

> ***ABZ***: This time you get to choose between pushing button A, pushing button B *and* doing nothing. Being a psycho makes you very likely to push A, and equally (and very) *un*likely to push B and to do nothing; not being one has the opposite effect. As before, pushing A or B causes all psychos to die. Pushing B also incurs the small cost Δ. So your situation is as in Table 3.4. As before, you are 95 per cent confident that you are *not* a psycho. What should you do?

Suppose that (3.3), (3.5) and (3.6) are all true, and consider first the choice between *A* and *Z*. The argument for preferring *Z* to *A* in *Psycho Button A* was that although *A* is unlikely to cause a bad outcome it *is a sign* that it will. But this is also a feature of *ABZ*: pushing button A is unlikely to kill you but is symptomatic of a causal arrangement in which it does. So if (3.3) is right then the same reasoning applies here too.

It doesn't follow that *Z* is rationally preferable to *A* in *ABZ*: *Z* may still be inferior to *B*, in which case both *A* and *Z* are utterly irrational – neither is rationally preferable to the other. Still, it does follow that whether or not *Z* comes out on top, *A* certainly does not. So *A* is rationally sub-optimal in *ABZ*.

It is equally clear that the case for preferring *A* to *B* is unaffected by the presence of an additional option *Z*. In *A or B* that case was that the dominance principle applies, given that your choice is causally irrelevant to

---

[7] See the discussion at Chapter 1 n. 14 above.

whether you are psychopathic. This remains true in *ABZ*. So by the same reasoning, *B* is not rationally optimal in *ABZ*.

Finally, the additional option *A* does nothing to prevent our transferring to *ABZ* the reason for preferring *B* to *Z* in *Psycho Button B*. It's still true that nothing in your choice between *B* and *Z* has any evidential or causal bearing on whether you are a psychopath. And your 95 per cent confidence that you are not a psychopath ensures that the risks plus the fixed cost of *B* do not outweigh its benefits. So *Z* is not rationally optimal in *ABZ*.

So from (3.3), (3.5) and (3.6) it follows that *none* of *A*, *B* and *Z* is rationally optimal in *ABZ*. This is not the innocuous conclusion that nothing is *uniquely* rational in *ABZ* (cf. Buridan's Ass). It is the catastrophic conclusion that *whatever* you do in *ABZ* is irrational: whichever option you take, some other option is (and always was) rationally preferable to it.[8]

It may be fair to say this of some agent whose beliefs and preferences are incoherent, or who faces infinitely many options. But this isn't a case of either of these sorts. Your beliefs in *ABZ* are, despite their science-fictional character, plainly sane and coherent. (Or if they are not sane, why not say the same about *Psycho Button A* and refuse for that reason to learn anything from it?) And there are not infinitely many options but only three.

We have two options. If we grant (3.4), and hence (3.3), (3.5) and (3.6), then we can accept Egan's intuition that CDT gives bad advice in *Psycho Button A*. But then Egan has understated the gravity of the situation. That in some cases CDT returns an irrational option is the least of our concerns. The real trouble is that in other cases there is no *rational* option. We must abandon the very idea that a rational agent can so much as choose from amongst finitely many options in as straightforward a scenario as *ABZ*.

The only – so in my view, the compelling – alternative is that at least one option in *ABZ* is rationally optimal. But this means that (3.3), (3.5) and (3.6) cannot all be true. And so (3.4), which motivates all of them, must fall. But then this objection to CDT fails altogether. If CDT is wrong, it isn't because it *misdirects* the agent's specifically causal information. At any rate, Egan–Gibbard cases do nothing to show this.

---

[8] The reader who is familiar with *Death in Damascus* (Gibbard and Harper 1978: 373) might think that CDT faces a similar difficulty in that case: whatever you decide to do, once you know that you are going to do it, CDT considers the alternative to be rationally preferable. (In fact CDT also creates this predicament in *Psycho Button A*.) Does this not mean that CDT reckons every option irrational there too? Not in the same sense. The difference is that in *Death in Damascus* there is always a CDT-optimal option for any *given* state of belief. Which option is CDT-optimal might change as the agent's beliefs evolve. This fact does raise *a* difficulty for CDT, which I discuss further at section 3.2 below. But it is compatible with the basic requirement that your available options are rationally comparable in the light of any particular pattern of beliefs. Egan's position on *ABZ* is not even compatible with that.

Table 3.5 Amalgamated statistics for button cases

	P. button A		P. button B		A or B?		ABZ		
	A	Z	B	Z	A	B	A	B	Z
P	198	2	100	100	198	2	198	1	1
¬P	2	3,798	1,900	1,900	2	3,798	2	1,899	1,899

I turn to two objections. First, I may seem to have left open whether *ABZ* preserves the credences that I stipulated in the three two-option situations. For instance, that you are only 5 per cent confident of being a psychopath on either option in *Psycho Button B*, does not entail that you have the same confidence in this proposition when option *A* is available. That is true but not relevant. We can stipulate a probabilistic structure for *ABZ* that retains whatever features of the two-option situations motivated (3.3), (3.5) and (3.6) in the first place.

Thus imagine that your credence function reflects the distribution in Table 3.5 of results for trials in which 16,000 subjects faced the four choice situations, 4,000 facing each one.

So for instance, in *Psycho Button A*, your confidence that you are a psychopath given that you push button A is 198 / (198 + 2) = 0.99. And in *ABZ*, your initial confidence that you are a psychopath is (198 + 1 + 1) / (198 + 1 + 1 + 2 + 1899 + 1899) = 0.05.

The reader may check that the distribution in Table 3.5 validates all of my other assumptions concerning your credences in the four cases. It is therefore coherent to stipulate credences on all four cases in a way that suffices for present purposes. In particular, if your credences for *ABZ* reflect the distribution in the three columns on the far right of the table then the arguments for (3.3), (3.5) and (3.6) are as applicable to it as they were to *Psycho Button A*, *Psycho Button B* and *A or B*.

You might still worry that no such sequence of trials can have taken place because if it had, then all the psychopaths would be dead by now. To get around this, we might interpret Table 3.5 simply as a model of credences rather than a description of real trials, since the coherence of the associated distribution does not require that it reflect the actual relative frequencies of anything. Alternatively we could imagine that the table reflects actual trials in which the device is set to 'stun' rather than to 'kill', and so can be used repeatedly, and that your values for outcomes are the same on either setting. This *is* a little far-fetched. But if you were willing to swallow *Psycho Button A* in the first place, then to object at this point on that score would perhaps be straining at a gnat.

The second objection is that even if your credences are as stipulated, CDT does not recommend pushing button A in *Psycho Button A*, as Egan and I are claiming.[9] The reason is that Causal Decision Theory should not be understood as a theory of what to *do* but rather as a theory of how to *evaluate* your options in the light of the information that you then have. In particular, given your credences at the outset of *Psycho Button A*, CDT might initially evaluate *A* over *Z*, but it doesn't then recommend that you *press* button A. The step from evaluation to action should respect the following principle:

(3.7) **full information.** You should act on your time-*t* utility assessments only if those assessments are based on beliefs that incorporate all the evidence that is both freely available to you at *t* and relevant to the question about what your options are likely to cause.[10]

Axiom (3.7) is incompatible with your rationally pushing the button at the outset of *Psycho Button A*, because at that point you do *not* have Full Information about the effects of doing so. The reason is that further deliberation is likely to give you more information on this point. In particular, learning that your $U(A) > U(Z)$ should raise your confidence that you will in fact push button A, at least on the assumptions (i) that you are aiming to maximize utility and not news value and (ii) that you know this. This in turn should raise your confidence that you are a psychopath and so also that pushing the button will cause your own death. But this should now make you more *reluctant* to push the button and so less confident that you will. It follows that CDT does *not* advise you to realize *A* in *Psycho Button A*, at least not at the outset.

I'll make three points about this. First: if CDT so understood does *not* endorse pushing button A in *Psycho Button A* then it faces the same difficulty that this argument raises against (3.4). That is, it recommends *Z* over *A* in *Psycho Button A*, *A* over *B* in *A or B* (which is a Newcomb problem) and *B* over *Z* in *Psycho Button B* (in which the full-information principle plays no role). But this leads to the same problem of non-transitivity that arose in connection with *ABZ*. So CDT, or in fact any theory that advocates two-boxing in the Newcomb problem, had better endorse pressing button A in *Psycho Button A*.

Second: even if this version of CDT does not advise you to push button A at the outset, deliberation may ultimately settle down to a state that endorses this option. In fact, on natural assumptions about how your beliefs about what you will do respond to the relative *U*-scores of the options,

9 Joyce 2012.   10 Joyce 2012: 127.

deliberation *could* only settle down at a state in which CDT is indifferent between A and Z and in which, therefore, it *does* endorse pushing the button.[11] And if it does not settle down then (3.7) is worse than useless, leading as it then does to perpetually oscillating intentions. So constraining CDT by (3.7) does nothing to change the situation: it would still be both necessary and possible to run something like my own argument against anyone who rejects that endorsement on the basis of (3.4).[12]

Third: if the agent knows from the start that he respects (3.7) then it is not even clear that the intellectual trajectory I just described can even get started. At the outset he has $U(A) > U(Z)$; but this should only increase his $Cr(A)$ if he already thinks that acts on his present utility recommendations alone. But he doesn't think that: he knows, on the contrary, that his now holding $U(A) > U(Z)$ is *not* by itself a reason for him to push the button: for all he knows, $U(A)$ and $U(Z)$ will evolve before he reaches the point of action. So if he knowingly follows (3.7) then the fact that now, at the outset, has $U(A) > U(Z)$ need not be evidence that he will ultimately push the button, and so not evidence that he is a psychopath.

We are not yet done with *Psycho Button A*, which illustrates a second causalist objection to CDT: that it endorses foreseeably regrettable options.

3.2   The Piaf maxim

F. Arntzenius has proposed a constraint on decision theory that he calls *Piaf's maxim* (PM): 'a rational person should never be able to foresee that she will regret her decisions'.[13]

It is easy enough to see that CDT violates PM in *Psycho Button A*. First let us define regret. I'll say that an agent *regrets* an option O if, after deciding on it, she believes that if she *had* taken some other option then the outcome *would* have had a higher *ex ante* news value than the present (*ex post*) expected value of the outcome.[14] It follows that once you have decided

---

[11] Joyce 2012: 134. In this case we have such an equilibrium if $Cr(P) = 0.1$ i.e. it is 9 to 1 that you are *not* a psychopath. See further the discussion of Deliberational CDT at section 3.2 below.

[12] Note that (3.7) does nothing to affect the argument that if (3.3), (3.5) and (3.6) are all true then none of A, B and Z is rationally optimal in *ABZ*. For instance, B cannot be rationally optimal: however confident you become that you will realize A, nothing that this reveals about the causal structure of the problem will make B look better; conversely, however confident you become that you will realize B, A will *always* look better.

[13] Arntzenius 2008: 277.

[14] Two points of clarification. (i) My understanding of the counterfactual operator $>$ generalizes Edgington's (2004: 21): your *ex post* confidence that you *would* have got an outcome with value n if you *had* realized some actually unrealized option O' is your *ex post* expectation of the chance, at the time of decision, of an outcome of *ex ante* value n conditional on: (a) your then taking option

to push button A (as CDT demands, or at least permits, in *Psycho Button A*), you will regret it, since you will then think that you are almost certainly a psychopath and so almost certainly would have been better off (i.e. not about to die) had you not pushed the button. This regret is foreseeable: so CDT violates the Piaf maxim.[15]

The solution as Arntzenius sees it is not to prefer EDT to CDT – that would be pointless, since EDT violates the Piaf maxim in Newcomb's problem. It is rather to adopt a new **Deliberational CDT** that bears a close resemblance to the 'full-information' version of CDT as discussed at the end of the last section. The main difference is that at *no* point does Arntzenius's theory actually tell you to do anything. It only tells you what it is rational to *believe* that you will do. More precisely, it tells you what degrees of belief about what you will do are stable, in the sense that if you have *those* beliefs, U-maximization gives you no reason to incline further towards any one of your options.[16] For instance in *Psycho Button A*, the deliberational theory does not endorse any one option, but it does identify an 'equilibrium' in which you have $Cr(P) = 0.1$ and $Cr(\neg P) = 0.9$. Assuming that 99 per cent of psychopaths and 1 per cent of non-psychopaths would push the button, this means that at that equilibrium your confidence that you will push the button is about 0.1. Clearly this theory gives a special place in decision theory to causal beliefs, because they are what determine whether your belief-state concerning your actions is stable. Equally clearly this is not the same as in the standard version of CDT.

There are two objections to Deliberational CDT. First, there should seem to be something very odd about the equilibria in which rational agents are supposed typically to find themselves. For instance, in *Psycho Button A*, the agent in equilibrium is supposed to be both completely indifferent between pressing the button and not, and yet also 90 per cent confident that he *will in fact not* press it. If you are in that state then when we ask 'What are you going to do?' you will say something like

---

O', and (b) all facts that are causally independent of O'. (ii) The reason for saying that the relevant counterfactual measures the *ex ante* value that the outcome would have had had O' been realized is that in general we cannot make sense of news values for options that are certain not to happen; formally this is reflected in the meaninglessness, on that hypothesis, of the conditional probabilities on the right-hand side of (2.8)(ii). Assuming that options are simple, letting Z be the (rich) set of *ex ante* possible outcomes and using V and V^* to measure *ex ante* and *ex post* values and Cr and Cr^* *ex ante* and *ex post* credences, this means that the availability of O' suffices for an agent's regretting O if and only if $Cr^*(O) = 1 \wedge \Sigma_{x \in V(Z)} \Sigma_{Z \in Z: V(Z) = x} \times Cr^*(O' > Z) > \Sigma_{Z \in Z: Cr(Z) > 0} V^*(Z) Cr^*(Z)$.

[15] For a more formal treatment (with slightly different numbers) see Arntzenius 2008: 291.
[16] Arntzenius 2008: 292.

this: 'I don't care what I do – I am just as willing to press the button as not; but, given that I am very probably not a psychopath, I will very probably end up *not* pressing.' What is odd about this answer is that you are citing a factor outside your control, namely whether or not you are a psychopath, as something that determines what you will do quite independently of your inclinations on the matter, which are supposed to be neutral.

It's not that prediction of your own future actions on the basis of a pre-existing state is *always* an evasion or denial of your own agency. A serially irresolute gambler might on one occasion be strongly inclined to give up and yet expect, and say, that it is very probable that he will enter the next casino that he passes. That *might* be meant to convey the expectation that the behaviour is compulsive and so not really up to him. But it could also be (and in this case probably is) a prediction, on the basis of his known character, about the future course *of his own deliberation*. What he is foreseeing is that whatever his present inclinations, actually going past a casino will make the excitement of cards, wheel and dice so vivid that he will *choose* to break the resolution.[17]

But *this* is not your situation in the Arntzenian deliberative equilibrium. We might imagine *Psycho Button A* as follows. Your hand is on a lever that you can push to the left or to the right when a buzzer goes off. Pushing it to the left depresses the button. Pushing it to the right surrenders the opportunity to depress it. The buzzer goes off in thirty seconds. You are completely indifferent between pushing it to the left and pushing it to the right. Your present equilibrium is a deliberative terminus: you know that nothing will happen in the next thirty seconds to make one option seem better than the other. And yet you are nearly certain ($Cr\,(Z) =$ 0.9) that when the buzzer goes off your hand *will* push the lever to the right, because you are nearly certain that some already fixed feature of your psychology will cause this to happen quite independently of your (neutral) inclinations. This amounts to *giving up* on your agency. An 'agent' in Arntzenian equilibrium is not really an agent at all, but rather as much a spectator of the impersonal working-out of his predetermined psychopathology as is anyone watching him.

The second objection sets aside the content of the deliberational theory and returns to the Piafian basis for preferring it to CDT in the first place. Why accept PM? Arntzenius gives an argument for the maxim that he explicitly intends to cover a slightly different version of it (which he calls

---

[17] For further discussion of this example see section 8.2 below.

'desire reflection'); but if that argument works at all then it vindicates PM.[18]

The argument is that a rational agent should not be open to a 'money pump'. A money pump is a finite sequence of offers each of which she accepts but whose net effect is a definite monetary loss, so that after accepting all of the offers she certainly has less money than if she had declined them all. Being open to a money pump is potentially disastrous, for by simple repetition of the sequence a cunning adversary could bankrupt the agent.

It looks as though an agent who violates PM is open to a money pump. Suppose that the agent accepts an offer *O* that she knows she will regret. Then the cunning adversary could offer her *O*, wait until she regrets it, and then offer her a small fee to revoke her acceptance. She is then certainly worse off than if she had declined the first offer:

> For instance, suppose that the default arrangement is that you get orange juice each morning from the stand outside your front door. But suppose that according to your current desirabilities you would rather get coffee than orange juice two days from now. Indeed, suppose that, other things being equal, you now suppose that it is worth 10¢ to get coffee rather than orange juice two days from now. But suppose that you also know that tomorrow you will think it more desirable to get orange juice rather than coffee the day after tomorrow, indeed that tomorrow, other things being equal, you will find the difference worth 10¢. Then it seems that the merchant at the stand can make money off you by having you pay 10¢ today to switch your order, for two days from now, to coffee, and then tomorrow have you pay another 10¢ to switch it back.[19]

Similarly we can construct a money pump against CDT. A self-aware follower of CDT facing *Psycho Button A* will make some choice at t_1. Whatever it is, she will pay ten cents at the later t_2 to reverse it. By the same reasoning she will pay a further ten cents at some still later time t_3 to reverse that reversal. (At least, this is so on the harmless assumption that

[18] The premise of a second argument (Arntzenius 2008: 279–80) is that doing something foreseeably regrettable is equivalent to doing something that you know a better-informed accomplice would advise you against doing, because we can treat your future self as just such an accomplice. As he correctly argues (2008: 282–95), standard Causal Decision Theory does not advise options that are foreseeably regrettable in *that* sense. For instance, if in *Psycho Button A* you are very confident that you are not a psychopath, then what you should think is that if you were to push button A then your future self would very probably think *wrongly* that you *are* a psychopath, and so could not count as better informed. The analogy between your future self and a better-informed accomplice does, however, cause trouble for EDT in the context of Newcomb's problem. I discuss this argument at section 7.4.2.

[19] Arntzenius 2008: 279.

there is a large enough temporal interval between her initial decision and the moment of truth.) She is then certainly twenty cents worse off than if she had declined both offers. And it looks as though CDT must take responsibility for this.

We can see why the argument fails once we try to get clearer on how the pump is supposed to work. In the example involving orange juice and coffee, the vendor first offers you the option of coffee (and not the default orange juice) in two days' time. But what exactly is he offering to sell you? Let it be Monday today: then here are two interpretations of what he says:

(3.8) If you pay me ten cents now then you will get coffee on Wednesday morning *whether or not* you change your mind in the meantime.

(3.9) If you pay me ten cents now then you will get coffee on Wednesday morning *as long as you don't* change your mind in the meantime.

Let it be clear to you which of (3.8) and (3.9) the vendor means. (There is no harm in supposing that. It is unsurprising and irrelevant that a credulous or confused agent stands to lose money to a vendor who is willing to lie to or to mislead her.) Then in either case there is no possibility of a money pump. Not on interpretation (3.8), because then you would have no opportunity to pay ten cents on Tuesday to reverse your acceptance of today's offer. And not on (3.9) either, because then you would not accept the offer in the first place, since you know that you would reverse the acceptance before you got the coffee.

The same goes for *Psycho Button A*. It may be that paying ten cents at t_2 reverses your initial decision *whatever you subsequently do*. Or it may be that paying ten cents at t_2 only reverses your initial decision *provided you don't reverse it back again at t_3*. Either way, there is no possibility of a money pump.

In short, a rational and self-aware follower of CDT will refuse to make the *revocable* decision that initiates the money pump because she knows that she *will* revoke it. But if the initial decision is *ir*revocable then there is no possibility of a money pump in any case.

Arntzenius seems to concede that there is a problem with the argument, at least on interpretation (3.9) of the vendor's offer. 'But', he writes, 'it still seems as if there is something deeply worrying about having such foreseeable switches in desires'.[20] But why? Such switches are perfectly ordinary and raise no clear difficulty. There is nothing *irrational* about the

---

[20] Arntzenius 2008: 279.

young man who, having to choose between a career as an artist and a career as a doctor, decides on the first on the basis of his present values, whilst fully aware that in twenty years' time he will regret it, since he will by then have adopted values that he now holds in contempt.[21] Of course at the later date he will *think* that he made a mistake. But that is because his later *desires* happen to be calmer than those of his youth, not on account of any superior *information*; and it is not even true, let alone a demand of rationality, that the horses of instruction are wiser than the tigers of wrath.

A fairer treatment of foreseeable regret in subjective decision theory would consider it a cost, but not necessarily a *prohibitive* cost as PM demands. That you will regret something should certainly weigh against doing it, but still it might be worth doing. And that weight itself will diminish in proportion to the rate at which you discount future goods and harms compared to present ones. In the limiting case you might e.g. weigh the immediate pleasure of a drug over the almost immediate regret that you would feel after taking it. Even that is not really irrational. That it seems so probably has something to do with the genuine irrationality that people who behave in this way often also exhibit.

But waive that point, and focus on the difference between (i) regret/relief that is based on partial information and (ii) regret/relief that is based on full information. It is certainly true that pressing the button in *Psycho Button A* will, you expect, cause you to feel regret of type (i). For you know that pressing the button will cause you to believe that you have pressed it, and hence to believe that you are probably a psychopath, at least just before you know what will become of you.

But you don't think it will cause you to feel regret of type (ii), at least not on the assumption that you are, *ex ante*, 95 per cent confident that you are not a psychopath. On the contrary, you should in that case be 95 per cent confident that the final effect of pressing the button is that you will have survived and all the psychopaths will be dead, so that at this point you will feel relieved or vindicated, not regretful. If its tendency to cause regret of type (i) counts against an action, its tendency to cause relief of type (ii) surely counts *more* strongly favour of it, at least from a causalist standpoint. So at least in this case, the defender of CDT has this further reason to deny that the anticipated regret caused by pressing the button counts decisively against doing so.

---

[21] Nagel 1970: 74 n. 1.

3.3   Objective Decision Theory

This book covers a dispute between two *subjective normative* decision theories. A theory is *subjective* if it describes or recommends an agent's choices on the basis of her psychological state rather than any external facts to which they might be answerable. It is *normative* if it aims not to say what real agents actually do but rather what they ought to do in some situation, whether that is specified subjectively or objectively. Both lines of criticism considered so far lie within the subjective normative tradition. But Mellor's objection is external to it. He claims that no plausible decision theory can be simultaneously subjective and normative.[22]

Perhaps the easiest way to introduce Mellor's objections to normative Subjective Decision Theory (SDT) is to explain its contrast with his own **Objective Decision Theory** (ODT). He himself puts it like this: according to SDT, a doctor is right to prescribe a drug if *she* thinks it will relieve the patient's pain, and the patient is right to take it because *he* thinks so too. But according to ODT the same thing makes both of them right, namely that it *will* relieve the pain.

More generally, what defines a decision situation for normative purposes is not what the agent thinks and feels about it but rather how things actually are. And an option's figure of merit is not the sum of its *news value* to you in various possible events, these being weighted by appropriate *subjective probabilities*. It is rather the sum of its objective value for you in various possible events, weighted by the *objective chances* that those events would have if you *were* to realize the option.[23]

Since chances are objective there is such a thing as getting them right. But 'getting them right' doesn't mean having true beliefs about their numerical values (although one could of course have these) but rather having a subjective credence that matches the chance.[24]

And what makes values objective is that there's such a thing as getting *them* right too, which is easy to see if you see how to get them *wrong*. The smoker might now think that cancer is no worse than giving up smoking. As he later discovers to his cost, it is much worse. I feared the dentist's drill because I thought the pain would be terrible. But in the event it was not

---

[22] Mellor 1983; Mellor 2005.   [23] Mellor 1983: 269–73.

[24] Mellor 1983: 274. In fact for Mellor one's subjective degrees of belief, though not the chances themselves, take interval and not point values: 'getting the chance right' must therefore really mean: having a credence interval in which the true chance lies. This complication won't affect what follows. (See ibid.: 274 n. 1.)

so bad.[25] In these cases the subject is in error about what Mellor calls the objective utility of some contemplated outcome, this being a misjudgement about how bad the outcome will really be at the time of its occurrence. Getting objective value right means getting *that* right.

The following case illustrates both departures from the subjective paradigm. You, having some unspecified, unserious malady, get the option to take a drug that has a chance of curing you: only you're not sure *what* chance. Studies show that the drug has an 80 per cent chance of working on one half of the population (group A) but only a 10 per cent chance of working on the other half (group B). Other than actually taking the drug you have no way of telling whether you are in A or B. Only 1 per cent of sufferers who do *not* take the drug are ever cured.

You also know that whether or not it works on them the drug causes flu-like side effects in all subjects. So if the drug works you suffer only the flu-like side effects. If it doesn't work you also suffer the symptoms of the malady. You've never had flu so you don't know how bad these will be. Should you take the drug?

We needn't go further into the numerical details to see how (i) errors about chance and (ii) errors about objective utility might in this case force SDT and ODT apart. (i) If you estimate correctly that the symptoms of the malady are slightly worse than the side effects of the drug without those symptoms, but are also (and wrongly) confident that you are in A, SDT (i.e. both EDT and CDT) will advise you to take the drug. But since really you are in B, this is objectively a bad idea. (ii) If you guess right that you are in B, but wrongly expect the symptoms of the malady to be much worse than the drug's side effects, SDT will again advise you to take the drug. Again you will have gone objectively wrong, although this time for a different reason.

So there is a material difference between Mellor's objectivist approach and the subjective theories studied here. Were we wrong all along, then, to have been looking for a normatively adequate subjective theory? Not necessarily. You could just reinterpret the 'disagreement' between subjective and objective theories as a case of talking about different things: what you subjectively ought to do (i.e. by your own lights), and what you objectively ought to do (if you like, by God's lights). What makes it objectively right to take an aspirin is one thing, i.e. that it works. What makes it subjectively right is something else, i.e. that you think it does.

[25] Mellor 2005: 139, 141.

But Mellor denies that there *is* any such thing as the subjective practical rationality at which SDT thinks it is aiming. He writes:

> To act rightly or sensibly according to [objectively] expected utility one need not have subjective utilities or probabilities. Nor does having them suffice. Suppose I do have them, only they're so wrong that I act [objectively] wrongly when I maximize my objective expected utility. Is my action really at all to be commended for conforming to that principle? Of course the subjective theory is descriptively 'as applicable to the deliberations of a monster as it is to that of a saint' ... but that doesn't mean we should at all commend monsters for being at least subjectively rational. But calling behaviour 'rational' without commending it just abuses the term, and concedes the subjective theory's prescriptive bankruptcy.[26]

But calling behaviour 'rational' without commending it needn't be an abuse of the term. To call behaviour subjectively rational is not to commend it, but to exculpate it from having gone wrong in a way that we should distinguish from the ways of going wrong that concern Mellor. Genghis Khan's desire to conquer the world was perhaps quite wrong. But given that he had it he went the right way about realizing it: That isn't to commend either the person or what he did. But it *is* to say that he didn't make any further mistake, i.e. of the sort that he would have been making if, having this desire, he had pursued some more timid or otherwise inept military strategy.

The analogy with inference is helpful in this connection. As well as forming true or false beliefs on the basis of perception or testimony, people also form them on the basis of other beliefs by applying inferential rules, for instance by inductive or deductive inference. Knowing from perception that Johnny wears a Stetson and from testimony that he owns a six-shooter, I might deductively infer what perhaps neither perception nor testimony could have told me, viz that somebody wears a Stetson and owns a six-shooter.

Clearly there is a sense in which this conclusion is correctly inferred even if its premises were not objects of knowledge but rather of false or irrational belief. Suppose I mistook Billy for Johnny. Or suppose my informant misled me. Still, by deducing that conclusion from those premises I have not made any *further* mistake, even if the conclusion is also false.

But just as we can distinguish the correctness of deductive reasoning from that of the beliefs on which it is based, so we can similarly distinguish

---

[26] Mellor 1983: 279.

the correctness of an agent's practical reasoning from the correctness of the credences and values on which *it* is based. This isn't yet to deny that there is also some objective sense in which both your actions and your subjective reasons for them can be called right or wrong, although I *will* shortly deny that. But it leaves room for a notion that stands to the objective correctness of one's credences and values as the validity of an argument stands to the truth of its premises.

But it's not just that ODT leaves room for SDT. Sometimes we need SDT to make distinctions that ODT obliterates. That happens when either (a) the agent's beliefs fail to track objective chances, or (b) the agent's desires fail to track objective utilities.

(a) You throw a die that is (everyone knows) heavily biased towards 2; it lands 6. So its present chance of landing 6 on that throw is 1. Its present chance of landing on any other number is 0. Alice and Bob know the bias but not the outcome, and each must bet on how it landed. Bearing in mind the bias, Alice bets that it landed 2. Bob, on a fancy, bets that it landed 4. Both are acting irrationally in the only sense that Mellor's ODT admits, since the objective utility of betting on 6 exceeds that of betting on 2 or on 4. But clearly there is a sense in which Bob was foolish whereas Alice was just unlucky. Only SDT can account for this.

Perhaps the objectivist will say that what rationalizes Alice's bet is not the present chance of its landing 6 (i.e. 1) but the chance of this *before* it was thrown. But then suppose that Charlie, who like Alice and Bob is betting on the outcome *ex post*, gets good evidence (say, the word of a trusted accomplice) that the die has landed 6. So he bets on 6. What rationalizes Alice's bet now makes Charlie's bet *ir*rational, since it is as true 'for' him as for anyone else that the actual *ex ante* chance of 2 was greater than that of 6. Clearly though, there is a sense in which Charlie was rational to bet on 6 *and* Alice was rational to bet on 2. Objectivism cannot account for this.[27] In fact only a subjectivist theory, that relativizes the rationality of

---

[27] Beebee and Papineau claim to identify a different sort of objective probability to which both Alice and Charlie are conforming their actions in this example. This is 'objective probability relative to the agent's knowledge of the set-up' (Beebee and Papineau 1997: 139). The objective probability of 2 is supposed to be high relative to Alice's knowledge. The objective probability of 6 is supposed to be high relative to Charlie's knowledge. Unfortunately they do not say much about how we are supposed to calculate these relative probabilities, other than that they are meant to follow from 'probabilistic laws', which can be known 'on the basis of inferences from statistical data about the proportions of *B*s observed in classes of *A*s' (ibid.: 140). But as Beebee and Papineau admit, these statistical laws are going to leave many cases wide open, cases in which agents must still act (ibid.: 142). So anyone who advocates this conception of objective probabilities must also admit, as they do admit, a distinct notion of subjective practical rationality (ibid.: 131–2).

an option to the agent's evidence, hence to his subjective beliefs, can make room for it.[28]

(b) The cases where it makes sense to speak of objective utility are the cases where it makes sense to say that one's preferences involve a mistake. For instance, it certainly sounds all right to say that I was wrong to dread the dentist's drill. There are two kinds of mistake that I might have made.

First, I might have been in error about how painful the drill was going to be, where 'pain' denotes an objective state of the nervous system whose duration and intensity are independent of the attitude that I took towards it either in advance or at the time of the drilling. (I am ignoring 'the power of positive thinking'.) Perhaps this error is a consequence of other false beliefs that I might have had: for instance, I might have thought that dentists use ordinary domestic power drills. Making a mistake about this would be like making a mistake about whether a drug causes drowsiness: the error consists in having a false belief or a skewed expectation.

Second, I might instead (or in addition) have been wrong about my own future attitude towards the drilling. For instance, I might have mistakenly thought that when I was under the drill I'd regret having gone under it. I might have expected that I'd *then* prefer the continuation of the toothache to the continuation of the drilling. Again, this false belief might be a consequence of other false beliefs; and again, what constitutes the error is a false belief that I later discover.

If my desire to miss the dental appointment is based on either of these errors then there is a clear sense in which extracting that desire makes for an objective improvement in my preference. But we can imagine *other* cases from which both sources of error are absent. In these other cases it seems possible for different persons to have opposite and yet objectively unimprovable preferences.

For instance, Alice and Bob are smokers. They are now (mid-December) contemplating quitting in the New Year. The only way to quit is to take a certain drug that is available on prescription until early January but will be unaffordable after that. The drug works by making smoking impossible

---

[28] This may be a sensible place to emphasize that 'subjectivist' labels what Mellor is rejecting, i.e. any view that entails that two people in the same type of situation with the same aims might act rationally by acting differently, in particular if their evidence is different. But 'subjectivism' in that sense is compatible with at least one 'objectivist' view, namely Objective Bayesianism (Williamson 2008), about which the only thing that matters here is that it imposes objective constraints (e.g. 'maximum entropy') on an agent's credential distribution given her evidence. That constraint is compatible with what I am calling Subjective Decision Theory and in particular with both EDT and CDT, although nothing in what follows presupposes it.

(because it causes an allergic reaction to tobacco). There is an unavoidable three-week period of withdrawal during which the quitter regrets having taken the drug. After that the ex-smoker is pleased to have given up. Smokers who do not take this opportunity will certainly regret it after the three-week period is up. Alice and Bob know all this. And neither of them is under any illusion as to how unpleasant the withdrawal symptoms would be.

Then it might be Alice's present preference to take the drug and Bob's present preference not to take it. In having these preferences *neither* of them is being irrational. For instance, it is no use pointing out to Bob that if he doesn't take the drug he will regret it. He already knows that (and in any case if he does take it he'll regret that too – only sooner). And it is not as though either party's future self knows anything about which his or her present self is in error. As I said, Alice and Bob both know perfectly well what withdrawal feels like. (Maybe both have tried to quit before.) So neither party commits either of the sorts of error that might have underwritten my, *genuinely* misplaced, fear of the dentist's drill. It is just that they are taking different present attitudes towards a future or counterfactual prospect that both of them see steadily and whole.

Alice and Bob therefore seem to have present preferences that are opposed and objectively unimprovable. So there cannot *be* any objective utilities in this situation. If there were, then at least one of Alice and Bob would be wrong to have the preferences that they do have; but neither of them *is* wrong. But this means that ODT says *nothing* about the rationality of the options that are open to Alice and Bob. How much should Alice be willing to pay to take the drug? How much should Bob be willing to pay to avoid it? It is plausible that Alice would be more rational to pay a low price than a high price to take the drug; also that there is an N such that Bob would be irrational to pay more than $\$N$ not to take it. But these estimates of rationality are simply unavailable to ODT. In this decision situation, and any other of this very common type, only a subjective decision theory can give us any idea of what an agent should or shouldn't do.

The burden of this section has been that Mellor's objectivism need not exclude a sense of practical rationality that subjective decision theory measures, the need for which is especially acute in cases where ODT delivers either (a) an implausible answer or (b) none at all. So decision theory *can* be both normative and subjective: that is, it can measure the rationality of an option against the fixed background of the subject's beliefs and desires.

And the burden of sections 3.1 and 3.2 is that *if* the subject's causal beliefs make a difference to that measurement then they do so in the way that CDT specifies. Whether they *do* make that difference therefore turns on the motivation for CDT itself, and in particular on the reasons, to which I now turn, for preferring it to the Bayesian alternative.

CHAPTER 4

Realistic cases

The simplest reason for preferring CDT to EDT is that only CDT gives the right recommendation in Newcomb's problem. But in its standard version (which covers the presentation at section 2.5), Newcomb's problem involves a science-fictional situation that never has faced and never will face any real person.

This makes the case problematic for three of the following four reasons. First, even if it *were* true that everyone's intuition in the Newcomb problem is that taking both boxes is uniquely rational, that need not constitute especially strong evidence that CDT gets this case right, since nobody has actually faced the situation and only a few people have given it any thought. Second, one might claim that our concept of instrumental rationality is just not specific enough to settle cases as esoteric as Newcomb's problem, for the simple reason that it has never needed to be. Third, the fact that people's intuitions do in fact *clash* over the Newcomb problem again gives us no particular reason to take seriously anybody's intuition in favour of two-boxing. Fourth, even if we concede that EDT gets the case wrong, the fantastical nature of the case means that it just doesn't *matter* that EDT gets them wrong.

Anyone who took any of these lines would not see in the standard version of Newcomb's problem much reason to prefer CDT to EDT, and this would be positive grounds *for* evidentialism. So the proponent of CDT must do one of two things. He must do *either*:

(a) (i) present more realistic examples of decision situations over which CDT and EDT disagree *and* (ii) show that CDT makes a better recommendation than EDT in those cases; *or*
(b) (i) counter these reasons for dismissing the standard Newcomb problem *and* (ii) show that CDT does better than EDT in *that* case.

This and the next two chapters cover candidate examples that might figure in an (a)-type argument, and Chapter 7 assesses the prospects for a (b)-type argument. More specifically, this chapter argues (secions 4.1–4.5) that all but one of the main examples cited in support of (a) in fact fail condition (a)(i): they are cases where EDT and CDT agree. And in the one case of disagreement that *may* be realistic, it is just not clear that CDT does better (section 4.6). Whether or not it does turns ultimately on the arguments that apply equally to the standard, fantastical Newcomb case, and so I defer judgement on that until Chapter 7.

Chapters 5 and 6 are a sort of counterattack. They present three examples of (a)(i) that fail condition (a)(ii): *relatively* realistic cases that *only EDT* gets right. Section 7.1 considers the 'science fiction' arguments at greater length and concedes that (b)(i) may be feasible. But the rest of Chapter 7 argues that (b)(ii) is not. So even if the standard Newcomb problem is relevant to the EDT/CDT dispute, it counts in favour of EDT. So too does the one realistic case that the present chapter leaves outstanding.

So in every case the agent *need* not, and in some cases he had *better* not, act on his causal beliefs at the expense of his evidential ones. That establishes premise (2) of my main argument as outlined at the top of Chapter 3.

4.1 Remedial cases

In the simplest realistic cases, which are not Newcomb-like, Evidential Decision Theory is alleged to prohibit certain *precautions* or *remedies* because they are themselves evidence of the ills that they are supposed to forestall or mitigate:

> [According to EDT] [p]atients should avoid going to the doctor to reduce the probability that one is seriously ill; workers should never hurry to work, to reduce the probability of having overslept; students should not prepare for exams, lest this would prove them behind in their studies; and so on. In short, all remedial actions should be banished lest they increase the probability that a remedy is indeed needed.[1]

Suppose you must choose whether to make an appointment with your doctor, Dr Foster. Either you are already ill or you are not. Whatever you do you prefer not being ill to being ill. But you prefer visiting the doctor (which carries an opportunity cost) if and only if you *are* ill. These possible options, events, and your notional dollar values for their combinations are as in Table 4.1.

It obviously can be sensible to visit the doctor even if you only suspect that you are ill – in effect it is a ten-dollar insurance against a 500-dollar loss.

---

[1] Pearl 2000: 108–9; see also Skyrms 1980: 130; Skyrms 1982: 700 ('Uncle Albert').

Table 4.1 *Dr Foster*

	S_1: you are ill	S_2: you are well
O_1: visit doctor	−500	0
O_2: don't visit doctor	−1000	10

And as long as your visiting the doctor *makes* no difference to whether you are ill (as surely it does not), CDT concurs. But EDT would advise against it if you take such a visit to be evidence that you are ill. And why shouldn't that be so? After all, most people only visit doctors when they *are* ill, so if *you* visit the doctor then this is a sign that *you* are ill.[2]

The case is evidently realistic. There certainly is a statistical correlation between visiting the doctor and being ill: you would e.g. find a greater

---

[2] If you want a more quantitative treatment, assume linear news value for dollar receipts. Suppose that 60 per cent of visitors to this doctor in this period are ill and 5 per cent of eligible non-visitors are ill. Then if your conditional credences reflect these frequencies, the V-scores of the options are as follows:

(i) $V(O_1) = -500 \, Cr(S_1|O_1) = -300$

(ii) $V(O_2) = -1000 \, Cr(S_1|O_2) + 10 \, Cr(S_2|O_2) = -40.5$

So $V(O_2) > V(O_1)$ and EDT recommends not visiting the doctor. Was it legitimate for me to assume linear news value for dollar receipts? You might think not, on the basis that linearity implies risk-neutrality, whereas people are generally risk-*averse* when it comes to serious illness (or thousand-dollar losses): see Chapter 2 n. 14. The assumption *is* unrealistic. But in this context it is harmless. Consider *any* distribution of values that preserves the ordering in Table 4.1, which abbreviating $V(O_i \wedge S_j)$ as V_{ij} we may write as follows:

(iii) $V_{21} < V_{11} < V_{12} < V_{22}$

Proposition (iii) and the Archimedean property of real numbers ensures the possibility of statistics that may reflect no *causal* influence *from* visits *to* illness but which, if themselves reflected in your credences, entail:

(iv) $V_{21} \, Cr(S_1|O_2) + V_{22} \, Cr(S_2|O_2) > V_{11} \, Cr(S_1|O_1) + V_{12} \, Cr(S_2|O_1)$

Proposition (iv) entails that EDT prefers not visiting to visiting. It is also compatible with unconditional credences about your health $Cr(S_1)$ and $Cr(S_2)$ that make it both intuitively rational and CDT-rational to visit Dr Foster. For instance if we choose positive Δ close enough to zero we can assign:

(v) $Cr(S_1|O_1) = 1 - \Delta$

(vi) $Cr(S_1|O_2) = \Delta$

(vii) $Cr(S_1) = \Delta + \{[V_{22} - V_{12}] / [(V_{22} - V_{12}) + (V_{11} - V_{21})]\}$

The reader will easily verify that for small enough Δ, (iii), (v) and (vi) guarantee that (iv) holds, whilst (iii) and (vii) together entail that $\Delta < Cr(S_1) < 1 - \Delta$. Finally, it also follows from (vii) and $\Delta > 0$ that:

(viii) $Cr(S_1) \, V_{11} + Cr(S_2) \, V_{12} > Cr(S_1) \, V_{21} + Cr(S_2) \, V_{22}$

And I am taking it that (viii) expresses an intuitively sufficient condition for its being rational to visit the doctor. For (viii) weights the news value of each outcome V_{ij} by the unconditional probability of the associated S_j; and this is what looks rational if, as in this case, whether you visit the doctor has no effect upon whether you are ill. Certainly in that case (viii) is a sufficient condition for CDT to recommend visiting the doctor.

concentration of genuinely ill people inside doctor's waiting rooms than outside them. Nor is this surprising. After all, people who are ill usually have symptoms. And people are likely to visit a doctor when they have symptoms. On the other hand there is not normally any causal connection between visits to the doctor and being ill – at least not from the former to the latter, setting aside illnesses that people tend to get *from* a medical environment.

A further obvious contrast with Newcomb's problem is that in *Dr Foster* there is no dominant option relative to the partition $\{S_1, S_2\}$. If you are well you are better off not visiting but if you are ill you are better off visiting.

A third difference is that your choice between O_1 and O_2 is evidence of the event (ill or well) because the event causes your act via *beliefs* about the former that motivate the latter. That is why people who never notice their symptoms are no more likely to make doctor's appointments when ill than when well. The involvement of the agent's beliefs at this point is common to all of Pearl's examples. Hurrying to work is a sign that you have overslept because oversleeping is something you can remember. Cramming for exams is evidence of being behind with your studies because people who are behind generally have some idea that they are. Nothing like this sustains the evidential connection in the standard Newcomb problem: even by its standards it is unrealistic to suppose that anyone would decline the transparent box because she thought she had been predicted to decline it.

But on closer inspection it is less clear that EDT and CDT really do disagree over *Dr Foster*, because it's not clear that EDT recommends not visiting. The argument that this was so turned on the inference, from the undisputed premise that there is a statistical correlation between visiting the doctor and being ill, to the conclusion that your conditional *credences* about your *own* case reflect this. It's true that people are more often ill when they visit the doctor than when they do not. But why should it follow that *your* visiting the doctor *now* is evidence for *you* that *you* are now ill?

Well, one might say, our credences just *do* tend to reflect the available statistical evidence (excluding grue-like contrivances). If I know that coin A has landed heads on 800 of its last 1,000 tosses, and that coin B has landed heads on 100 of its last 1,000 tosses, then I will factor this into my conditional credences concerning the next toss of one of these coins. I will, for instance, take the news that the next toss is of coin A to confirm the hypothesis that the result in this single case will be heads. We also apply the statistics behind *Dr Foster* when estimating the health of *others*. If you learn that Jones has made an appointment to see Dr Foster then this evidence will raise your confidence that *he*, Jones, is ill. So of course, one might continue, the same is true in the *first*-personal case. If you learn that you have just made an appointment to see Dr Foster then you will become more confident that *you* are ill. Or if you don't then you should. Or do you consider yourself above the law?

But that is manifestly not what we do. Any ordinary, apparently rational person who makes a doctor's appointment does *not* then become *more* confident that he is ill than he was before picking up the telephone. If that were true, then we should also observe the following: somebody who is reasonably confident that he is ill will call the doctor to make a regular appointment. After putting down the telephone he becomes *more* confident that he is ill. But this increased confidence raises his motivation to take remedial measures (remember that that is what explains the statistical correlation between illness and doctor's appointments in the first place). So now he calls again to make an emergency appointment, or visits the pharmacy, or starts preparing a will, etc.

Similarly, if it is true in general that agents count their own remedial options as evidence of what they are remedies against, then we should expect remedies to be self-reinforcing: the more remedies one takes, the more one is inclined to take remedies. In fact if Pearl was right about this, then we should expect not only remedies but also *any* options that are based on evidence to be self-reinforcing. Your hurrying to work would be evidence that you are late for work, and since people do in fact hurry more as they become more confident that they are late (in accordance with CDT), we should expect someone who starts to hurry because he thinks he might be late to become more confident that he is late because he is now hurrying, and so for that reason alone to go faster, or to take a taxi. Again, your buying shares at a certain price would be evidence for you that they will do well. But once you have got that evidence, you should bid for more at a higher price.

But these things do not happen. A commuter does not take his running to work as a further reason in itself to go faster. Buying shares at a given price is not in itself a reason to buy more at a higher price. Nobody takes his own remedial options to be evidence of the events that they are remedies against.[3]

But is this not something that we *should* be doing? After all, there *is* a statistical correlation between (say) visiting Dr Foster and being ill. You apply this correlation to draw conclusions about *another's* state of health from the evidence that he has made an appointment. And others will apply it to draw similar conclusions about *you*. So isn't it just an error on your own part not to draw this conclusion about yourself? If so, Pearl's argument still

---

[3] For counterexamples see e.g. Knox and Inkster 1968: this makes no difference to the normative point to follow.

threatens EDT. It isn't much comfort to EDT that only our own epistemic irrationality saves it from advising us against remedies.

A promising line of reply is that what both explains and justifies your not applying the statistical correlation to yourself is that you already have more evidence about yourself than you do about others. This additional evidence renders your appointment with Dr Foster evidentially irrelevant to your state of health. In the jargon, it *screens off* the one from the other.

In particular, your background evidence E might include a detailed description of your own symptoms. And probably amongst people with symptoms fitting that description, there is *no* statistical correlation between visits to the doctor and illness. A person with just those symptoms, e.g. you, is therefore in your estimation just as likely to be ill whether he visits the doctor or not.

Slightly more formally, write S_I^{JONES} and S_I^{YOU} for the propositions that Jones is ill and that you are ill, respectively. Similarly let O_I^{JONES}, O_I^{YOU}, O_2^{JONES} and O_2^{YOU} respectively express the propositions that Jones visits the doctor, that you do, that Jones doesn't visit the doctor, and that you don't. And write E for *your* background evidence concerning both Jones and yourself. Then your doxastic position is:

(4.1) $Cr(S_I^{JONES}|O_I^{JONES}) = Cr(S_I^{JONES}|O_I^{JONES} \wedge E) > Cr(S_I^{JONES}|E)$
$= Cr(S_I^{JONES})$

(4.2) $Cr(S_I^{JONES}|O_2^{JONES}) = Cr(S_I^{JONES}|O_2^{JONES} \wedge E) < Cr(S_I^{JONES}|E)$
$= Cr(S_I^{JONES})$

(4.3) $Cr(S_I^{YOU}|O_I^{YOU}) = Cr(S_I^{YOU}|O_I^{YOU} \wedge E) = Cr(S_I^{YOU}|E) = Cr(S_I^{YOU})$

(4.4) $Cr(S_I^{YOU}|O_2^{YOU}) = Cr(S_I^{YOU}|O_2^{YOU} \wedge E) = Cr(S_I^{YOU}|E) = Cr(S_I^{YOU})$

What (4.1)–(4.4) are saying is that in light of your meagre knowledge of Jones's symptoms, you take *Jones's* visiting the doctor to be evidence that *he* is ill. But in light of your much fuller knowledge of your own symptoms, you don't take *your* visiting the doctor to be evidence that *you* are ill. Equivalently, E doesn't screen off S_I^{JONES} from O_I^{JONES}. But E *does* screen off S_I^{YOU} from O_I^{YOU}. And this asymmetry of background knowledge both explains and justifies your willingness to apply the statistical generalization to Jones but not to yourself.

But this cannot be the whole story. It is easy to imagine cases where your background evidence concerning yourself is no better in any relevant respect than your background knowledge concerning Jones. For instance, suppose you know that you and Jones awoke at the same time (the same alarm clock aroused both of you), but you cannot remember what time that was. So you have no better idea of whether you have overslept than of whether Jones overslept. So there is nothing to distinguish Jones's act from yours. Your background evidence cannot make your own hurrying to work evidentially irrelevant to whether you overslept if it doesn't make Jones's hurrying to work evidentially irrelevant to whether *he* overslept. And yet the asymmetry still exists: if you see *Jones* hurrying then you will become more confident that *he* (and you) overslept; but if *you* hurry then that won't make you any more confident that *you* (or he) overslept. So notwithstanding this 'promising line of reply' to him, Pearl could say that in this case you *would* be rational to take your own hurrying to work as evidence that you are late. Here, then, is a remedy that EDT seems wrongly to advise you not to take.

A better defence of the asymmetry should apply even to this case. Consider a similar asymmetry that arises for testimony, by which I mean verbal expressions of (varying degrees of) confidence in some proposition. *Another person's* testimony can, though it needn't, have evidential bearing for you upon its subject matter. But *your own* present testimony could not have this bearing for you. Jones's saying 'I think it's going to rain' might increase your confidence that it's going to rain. And if he says 'It certainly will rain' then that will increase it more. But *your* asserting it with any degree of confidence has no such effect on your own confidence of this, although it might of course affect Jones's.

This is hardly surprising. When you testify that *p*, your act of testimony already takes account of all and only the evidence for *p* that you already presently possess. To take your own act of testimony as further evidence of *p* would be to double-count that evidence. But *Jones's* testimony that *p* is evidence that Jones has evidence that *p* that (for all you know) you might *not* possess. So it is entirely rational to raise your confidence in *p* in the second case, but not in the first.[4]

---

[4] There may be cases where you are rational to amend your beliefs in the light of Jones's testimony even though you *know* that you and Jones have the same evidence. For instance, suppose you and Jones are looking at a large crowd of people from the same perspective and are asked to judge the size of the crowd, you and Jones being acknowledged peers at this sort of estimate. And suppose you think that your evidence is *unbiased*, in the sense that judgements of peers with this visual evidence are randomly and symmetrically distributed about the true value, the variance of the distribution being inversely proportional to your common competence at such judgements. Then Jones's answer should, if it differs from yours, cause you to amend your initial judgement. But the explanation for this is that on the hypothesis that you are equally good judges, Jones's testimony itself adds to your own evidence about the case, so taking account of it does not involve double-counting any of the evidence that prompted your own initial judgement. For this reason the argument in the main

And what goes for verbal expressions of confidence goes also for those non-verbal expressions of confidence that remedial actions constitute. These are all a kind of testimony in the sense that one's beliefs speak through them. (i) Your visiting the doctor evinces a degree of confidence that you are ill. (ii) Jones's hurrying to work evinces a degree of confidence that he overslept. But this only makes them *evidence* of those events (being ill, having overslept) to persons who (think they might) lack your evidence in case (i) or Jones's in case (ii): so *not* you in (i), or Jones in (ii).

More generally, remedial actions are evidence of what they are remedies against only because they evince a perspective from which that event seems likely. But this is only evidence of the event to one who does not already see things from that perspective. If you know that Jones remembers whether he overslept and you do not, then his hurrying to work is evidence for you that he did. But if you know that you and Jones remember things just the same, perhaps because you *are* Jones, then this no longer follows.

For this reason not only is it factually accurate but it also imputes no error to us to say that each of us does in fact take her own remedial actions to be *evidentially* (as well as causally) irrelevant to the events that they are remedies against. So a rational as well as plausible specification of your subjective beliefs Cr in Dr *Foster* conforms to the following:

(4.5) $Cr\,(S_1|O_1) = Cr\,(S_1|O_2) = Cr\,(S_1)$
(4.6) $Cr\,(S_2|O_1) = Cr\,(S_2|O_2) = Cr\,(S_2)$

It follows from (2.18), (4.5), (4.6) and Table 4.1 that EDT recommends visiting the doctor unless you are very confident that you are not ill (unless $Cr\,(S_1) < 0.02$, if news value equals dollar value). EDT here agrees with both common sense and CDT. So whilst they *are* realistic and straightforward, these remedial cases do not constitute any sort of reason for preferring EDT to CDT. EDT and CDT do not even disagree over them.

Apart from this conclusion, the most important point to take forward from this discussion is the distinction between statistical correlations (relative frequencies) and *subjective* evidential relations in the form of conditional credences. It is the latter, not the former, that drive Evidential Decision Theory. This makes a difference when, as here, they come apart. It may be the conflation of these things that made remedial cases seem to Pearl so clearly to refute EDT. Perhaps that also accounts for his strangely abusive and ill-informed remarks about that theory.[5]

text is in no obvious tension with a 'conciliationist' approach to peer disagreement (such as the 'equal-weight' view – see Elga 2007: 484 ff.).
5   Pearl 2000: 108–9: it is a 'bizarre theory' that generates 'obvious fallacies'; it was 'a passing episode in the philosophical literature, and no philosopher today takes the original version of this theory

This is also an appropriate place to reject the canard, which may also arise from conflating these two things, that EDT recommends 'the ostrich's useless policy of managing the news'.[6] The association with ostriches is damaging: nobody thinks that hiding from your problems is going to solve them. So too is the association in many people's minds between 'managing the news' and the approach beloved of governments past, present and future, of publishing false or misleading statistics rather than tackling the problems that the true figures reveal. But neither of *these* things counts as news management of the *subjective* type that EDT would recommend. Not the ostrich's policy, because his not seeing the predator is hardly good news to the ostrich who knows that he is shutting his eyes to it. And not the manipulation of statistics, because whilst this may placate the common people (before cynicism sets in), it does nothing to reassure those persons in the Ministry of Truth who fabricated the figures in the first place.[7]

EDT does *not* recommend simply 'managing the news', at least not in so far as that implies misleading oneself or others. What it really recommends is *managing what is still news to you if you manage it*. It remains to be seen whether anything is wrong with that.

4.2 Medical cases

In response to the pressures mentioned at the start of this chapter, advocates of Causal Decision Theory have sought realistic analogues of the standard Newcomb case, i.e. ones that make no appeal to advanced alien technologies, beings with supernatural powers etc.

There is no clear a priori reason why our world should not contain Newcomb-like phenomena in abundance. What is needed is only that the agent's options be evidentially relevant but causally *ir*relevant to some partition of S (a) each member of which settles what matters to the agent given the choice of option, and (b) with respect to which an option that is evidence of a relatively good outcome be dominated by one that is evidence of a relatively bad outcome. It is not even necessary that the evidential relations be very tight (that the 'predictor' be perfect). As long as the options have *some* evidential bearing on the events, a big enough

---

seriously'. 'The original version' presumably means that in Jeffrey 1965, which is more or less the theory defended here.

[6] Lewis 1981b: 310.

[7] Nor, of course, does EDT recommend e.g. taking a drug that would make you irrationally optimistic. Whilst it's true that doing so would affect your perception of your own welfare, *ex ante* you do not think that the proposition that you take the drug is any sort of evidence that you really are or will be any better off.

difference between the values of the outcomes will ensure that EDT and CDT disagree.[8]

In one possible realization of this structure, both a particular option and some undesirable event are *effects* of a *common* cause. That option, which I'll call the symptomatic option, is a better *sign* of the undesirable event than its alternative, since effects are signs of their causes and so also of side effects of their causes. But it does nothing to *bring about* the undesirable event. If we could contrive that in both the undesirable event and also in its alternative, the outcome of the symptomatic option was preferable to the outcome of *its* non-symptomatic alternative, we should have a Newcomb case. Most 'realistic' versions of the Newcomb problem are of this general type.

Often the undesirable event is some medical condition that shares a common physiological cause with the symptomatic option. But although the undesirable event of which it is *symptomatic* is very bad, the net *effect* of the option is mildly positive. The following example is entirely representative.

> *Smoking Lesion*: Susan is debating whether (O_1) not to smoke or (O_2) to smoke. She believes that smoking is strongly correlated with lung cancer, but only because there is a common cause – a lesion that tends to cause both smoking and cancer. Once we fix the absence (S_1) or presence (S_2) of this lesion, there is no additional correlation between smoking and cancer. Susan prefers smoking without cancer to not smoking without cancer; and she prefers smoking with cancer to not smoking with cancer. Should Susan smoke? It seems clear that she should.[9]

This case does not describe a situation that anyone has actually faced. It is clear enough these days that smoking *is* a cause of lung cancer and does not just share causes with it. Nor was there *ever* any good reason to prefer the common-cause hypothesis.[10] Still, there may have been people who firmly believed that they faced Susan's situation. And since EDT and CDT are both *subjective* theories, i.e. advise the agent in the light of *her* beliefs and desires, these theories should apply to anyone who holds that false but sane and coherent belief.

And if we do apply them then it seems clear enough that they disagree for just the same reason that they disagree over the standard Newcomb problem. It is unnecessary to go into numerical details. CDT advises Susan

---

[8] Lewis 1979c: 302–3.

[9] Egan 2007: 94 with trivial variations. This example originally appeared in Stalnaker 1980 [1972]. For discussion of it see also Horgan 1981: 178–80; Eells 1982: 91; Skyrms 1984b: 65–6; Horwich 1987: 178–9; Edgington 2011: 77. For other medical cases see Nozick 1969: 218; Skyrms 1980: 128–9; Lewis 1981b: 310–11; Price 1986a: 196–8; 1991: 162–3; and 2012: 511–2; Mellor 1987: 170–1.

[10] For this point, and documentation of R. A. Fisher's *conjecture* of a common cause, see Seidenfeld 1984: 205–6; Levi 1985: 235–7.

to smoke because nothing that she does can prevent or bring about her having the lesion, and whether or not she does already have it, she is better off smoking. EDT advises Susan not to smoke because smoking is symptomatic of a greater evil than is non-smoking – assuming that, like most people, Susan very much prefers not smoking without cancer to smoking with cancer.

So this type of decision situation apparently could arise in real life. And it raises the same problems for EDT as the standard version of the Newcomb problem. In fact the situation for EDT is worse than that. Whereas there is no overwhelming consensus on the rational option in the standard Newcomb case, everyone agrees that Susan should smoke in *Smoking Lesion*. If Evidential Decision Theory disagrees with CDT over cases that are as realistic as this, then it is not only wrong but also dangerous.

4.3 The Tickle Defence

The standard evidentialist response is the 'Tickle Defence',[11] of which the most plausible versions share this basic structure:

(4.7) **Mental causation:** the undesirable event produces the dominant option via and only via its operation upon the agent's current pre-decision desires and beliefs.

(4.8) **Transparency:** the agent knows enough about her own current desires and beliefs to make the dominated option evidentially irrelevant to them.

(4.9) Therefore, the agent takes *neither* option to be evidentially relevant to the undesirable event.

The following exposition and defence of this argument focuses on *Smoking Lesion*, but generalizes to any realistic and allegedly Newcombian case in which the dominating option is symptomatic of some physiological cause of it.

4.3.1 The mental causation premise

This chapter is discussing cases that could actually occur in real life, not cases that we might just make up for philosophical purposes. If *Smoking Lesion* is to stand a chance of being of this type then we must be able to tell

---

[11] Horgan (1981: 178–9); Eells (1982: Chapter 7); Jeffrey (1983: 25); Horwich (1987: 179–85); and Price (1986a: 195–202; 1991; 2012: 511 ff.) all endorse versions of it. Skyrms (1980: 130–1) rejects a less plausible, phenomenological version. For additional discussion of Price see Ahmed 2005, of which remark (ii) in section 4.3.3 is a highly condensed version. The present version of the Tickle Defence incorporates elements of Jeffrey's, Horwich's and Price's versions.

some at least slightly plausible story about how smoking can be evidence of the lesion, and in particular about the mechanism by which the lesion causes the smoking.

Only two sorts of story are remotely realistic. According to the first story, what causes smoking is some causal chain that completely bypasses any brain-states or mental states that figure in its normal aetiology, such as the desire for a cigarette, or the presence, sensitivity, and dispersion in the brain of the acetylcholine receptors that regulate the desire for nicotine. Instead, the lesion makes you reach for a cigarette whether or not you desire one. Perhaps Tourette's syndrome and dystonia are models of this. There, the verbal or motor performance is more or less compulsive, occurring whether or not the speaker really desires it.

If *Smoking Lesion* were a case of that sort then Susan would be right to take compulsive smoking to be evidence of a pre-existing lesion and so to have $Cr(S_2|O_2) > Cr(S_2|O_1)$, just as she might take certain spasmodic movements of her hand to be evidence of a pre-existing neural condition like dystonia. But then there is no reason to think that compulsive smoking is an *option* for her, and so no reason to think EDT takes its being bad news (which it is) as a reason not to *do* it. As (2.18) makes clear, EDT pronounces on the rationality of *options*, i.e. propositions that are available for the agent to realize in some intuitive sense that it takes for granted. There is no intuitive sense in which compulsive smoking is available to Susan. Of course she *might* smoke *deliberately*. But on this hypothesis about the working of the lesion, deliberate smoking is evidentially irrelevant to the presence of the lesion, so is not bad news.[12]

---

[12] Sobel rejects a remark of Jeffrey's to the same effect (Jeffrey 1983: 25) on the grounds that 'to say that an action is caused, caused this way or that, directly or whatever, is not to say that it will take place quite regardless and that the agent has no choice in its connection. Consider that even if an action will be caused directly by some already-in-place condition (e.g. a subliminal stimulus), it is possible that it will also be chosen, and possible for both the in-place direct cause and the choice to be necessary causes as well as sufficient in each other's presence' (Sobel 1990: 44). But let the undesirable event S_2 be such an in-place condition. Then if it is already known by the agent to obtain, then nothing that she does could confirm it, i.e. $Cr(S_2|O_2) = Cr(S_2)$, and the case presents no difficulty for EDT. If S_2 is *not* already known to obtain then the possibility remains (epistemically) open to the agent that she fails to realize O_2, not because she chooses O_1 instead, but because she chooses O_2 in the absence of S_2. Then the agent's options are not $\{O_1, O_2\}$ – which is not a partition – but $\{O_1, \neg O_1 \land (S_2 \supset O_2)\}$. But then again, $Cr(S_2|\neg O_1 \land (S_2 \supset O_2)) = Cr(S_2)$ and the case presents no difficulty for EDT. For example, let O_1 be the proposition that I don't pull the trigger, let O_2 be the proposition that the gun goes off, and let S_2 be the proposition that the safety catch is off. Then S_2 meets Sobel's condition of being both causally necessary for O_2 and causally sufficient for it in the presence of my choice to realize $S_2 \supset O_2$, which here amounts to the proposition that I pull the trigger. But $Cr(S_2|\neg O_1 \land (S_2 \supset O_2)) = Cr(S_2)$: nobody thinks that pulling the trigger by itself gives me any more evidence that the safety catch is off than I already had before I pulled it. Nor would EDT advocate not pulling the trigger if for some reason I have an overriding interest in the safety catch's being engaged. (Similar remarks apply to van Fraassen's 'bumbling' version of the Newcomb problem: Jeffrey 1983: 20; see also Joyce 1999: 158–9.)

The more plausible hypothesis is that the lesion produces smoking via its effect on the agent's *desires* and *beliefs*. We are familiar with cases of this sort. For instance, real life offers plenty of cases in which prior states of the person affect his present desires. Smoking itself is thought to increase the desire to smoke by activating the mesolimbic or 'reward' pathway. Something similar may be true of many addictive drugs. Again, smoking may distort the beliefs that are motivationally relevant to smoking. For instance, it seems that smokers tend to underestimate the health risks to themselves of smoking.[13] However things actually are, it is therefore not wildly implausible that a lesion might cause smoking via effects on the smoker's desires and beliefs that are similar to those that smoking itself might have.

Note finally that if the lesion does cause smoking in any realistic such way then it does not cause smoking only amongst people who *know* themselves to be facing *Smoking Lesion*. People that have the lesion will, because of its presence, smoke more than those that don't have it, whether or not they have any idea that this is what causes them to smoke and whether or not they have any idea of what else it causes. If the lesion promotes some belief in Susan then what it promotes is the belief that smoking is not greatly injurious to her own health in general, not just some specific belief about the effects of her smoking on this particular occasion, or about what doing so on this particular occasion might indicate. Similarly, if the lesion promotes some desire in Susan, then what it promotes is the general attractiveness of smoking to her, not just an urge to smoke now that she is facing, and knows that she is facing, *Smoking Lesion*.

This is what makes it absurd to think that the lesion causes smoking in the only other way that it could cause smoking: that is, by promoting adherence to a decision rule. For instance, it would explain the correlation between having the lesion and smoking, in *Smoking Lesion*, that the lesion promotes adherence to Causal Decision Theory (at least on the supposition, which the case against EDT presupposes, that EDT promotes non-smoking). That would be absurd because if the case is at all realistic, the lesion must promote smoking in cases where CDT and EDT are not in dispute: thus as I said, the lesion equally promotes smoking amongst people who have no idea that smoking is evidence of its presence. It could only do this if it promotes smoking through some route that does not run via the agent's decision rule. If we are willing to engage in the fantasy of a physiological condition that promotes causalism then we might as well deal

---

[13] Arnett 2000.

directly with the standard Newcomb case. (This point does not apply to the *psychological* version of the Newcomb problem, which I treat separately at section 4.5.)

4.3.2   *The transparency premise*

The philosophers that have used the Tickle Defence of EDT are not always explicit about their grounds for premise (4.8). But two reasons look especially promising, of which I'll discuss one here and one at section 4.5.

The first is that it is simply a presupposition of normative Subjective Decision Theory that the agent knows perfectly what her own beliefs and desires are. This is because unless she knew this, the theory, which distinguishes rational from irrational options on the basis of given desires and beliefs, is not one that she could use to make a prescription at all. Thus e.g. Horwich writes,

> One should not object here that a person's desires may not always be accessible to introspection. This is true but irrelevant. [The Tickle Defence] needs to be employed only for situations that provide alleged counterexamples to the evidential principle. And there can be a counterexample to the principle only if the principle is applied, and therefore only if the belief and desires of the agent *are* known by him at the time of deliberation.[14]

If Horwich is right then the transparency premise follows. Suppose Susan knows for certain both her credences $Cr\,(S_1) = x_1$, $Cr\,(S_1|O_1) = x_2$ and $Cr\,(S_1|O_2) = x_3$, and her values $V_{11} = m_1$, $V_{12} = m_2$, $V_{21} = m_3$, $V_{22} = m_4$, here writing V_{ij} for $V\,(O_i \wedge S_j)$. Then her credences in each of *these* equations is 1, in the first three cases these being 'second-order' credences. But if your confidence in a proposition is 1 then no unknown proposition has *any* evidential relevance to it. In particular, her not smoking is evidentially irrelevant to her present beliefs and desires since in that case we have e.g. $Cr\,(Cr\,(S_1|O_1) = x|O_1) = Cr\,(Cr\,(S_1|O_1) = x)$ for any x.

But there are two objections to this argument. The first is that if applying the 'evidential principle' really does require perfect knowledge of one's own credences and values then the principle is probably inapplicable. Only the most extreme Cartesian view of 'our knowledge of the internal world' would allow that agents could identify these quantities to the nth decimal

---

[14] Horwich 1987: 183. See also Eells 1982: 158.

place by introspection. And no other means are realistically available for achieving perfect knowledge in this area.[15]

The second objection is that in any case, the agent's own use of subjective decision theory as a prescriptive instrument does *not* require *precise* knowledge of her beliefs and desires. This is most obvious in connection with the advice that *Causal* Decision Theory gives to Susan: *whatever* the precise magnitudes of her $Cr\,(S_1|O_1)$, $Cr\,(S_1|O_2)$ and $Cr\,(S_1)$, CDT advises her to smoke, and if she is going to use CDT then she ought to know this. In that case she needn't know anything at all about *those* values. As for her desires, it suffices for her to know that $V_{11} < V_{21}$ and $V_{12} < V_{22}$.

A similar point applies only slightly less straightforwardly to EDT. EDT advises Susan to refrain from smoking if and only if the following condition holds:

$$(4.10) \quad V_{11}\,Cr\,(S_1|O_1) + V_{12}\,Cr\,(S_2|O_2) > V_{21}\,Cr\,(S_1|O_2) + V_{22}\,Cr\,(S_2|O_2).$$

But she can know this without knowing the precise values of all these parameters. For instance if she knows that $V_{11} = 2$, $V_{12} = 0$, $V_{21} = 3$ and $V_{22} = 1$ (or that some positive affine transformation of V gives these values), then it suffices, for her to know (4.10), that she know only $Cr\,(S_1|O_1) >$ 0.75 and $Cr\,(S_1|O_2) < 0.25$. She can know that without having any more precise idea of what these credences are.

Generalizing the point, all that Susan or any agent needs to know about herself in order to apply CDT is that she takes this or that option to have the best net *effect*, as far as she is concerned. And all that she must know about herself in order to apply EDT is that she takes this or that option to be the best *news*, again as far as she is concerned. So Horwich is wrong to claim that the prescriptive use of these theories demands anything like the intimate knowledge of one's own mental state that his own version of the transparency premise entails.

Still, this second objection invites a reformulation of Horwich's point that avoids both difficulties. Let us suppose that Susan knows just the following about herself:

---

[15] Savage's theory does of course imply that the totality of one's preferences over acts uniquely determines one's subjective probabilities. So on that view it would be possible in principle to settle one's personal probability for any event in the field of $\succ$ with arbitrary accuracy, and one's utility function up to positive affine transformation, if one could settle one's actual or hypothetical preferences between arbitrary pairs of acts (see sections 1.4–1.5 above). But this procedure is hardly available in practice, since it involves all pairwise comparisons amongst the infinitely many functions that count as acts in Savage's definition.

(4.11)   That smoking is worse news for her than non-smoking.
(4.12)   That if she refrains from smoking then it will be for this reason.

That Susan knows (4.11) and (4.12) is not a very demanding assumption. Her knowledge of (4.11) requires no more detailed knowledge of her own credences and values than is implied by (4.10). And besides, we ordinarily credit agents with this much knowledge of their own news values: in fact it is something that we can hardly help presupposing when we bring our intuitions to bear upon this case. What you imagine, when you imagine being in Susan's position, is not only that smoking is bad news but also that you know this. To think of yourself as facing a Newcomb problem in which you are *ignorant* of your own preferences over options is none too easy a thing to do: in so far as we have any intuitions at all about *that* case (I have none) we should take them with a pinch of salt. This isn't to say that one could *never* learn about one's news preferences over the options in a decision situation only *ex post*; but it *is* to say that such cases are so unfamiliar as to give them no advantage, in respect of realism, over the standard, science-fictional version of the Newcomb problem.

Similarly with (4.12), it is certainly the normal situation that one knows when deliberating that: if one takes this option it will be for this reason and if one takes that option then it will be for that reason. If such competing reasons do exist and are known – and in this case they do and they are – then *not* to be able 'to explain oneself to oneself' in this way normally would, and I think should, undermine one's claim to the status of an agent. In particular, we can expect that anyone who knowingly faces *Smoking Lesion* will also know that she will refrain from smoking if and only if she does so because she takes it to be bad news. Probably it *can* happen that one thinks one might take this option for some subconscious and unknown reason, or that one thinks one might take that option for no reason at all. But the dialectical situation makes it reasonable to assume that Susan can rule out these possibilities. The point of *Smoking Lesion*, indeed of all versions of Newcomb's problem, is to show that the evidentialist goes wrong *because* she concerns herself only with the relative auspiciousness and not with the relative efficacy of her options. If it shows that then it should apply as well to an evidentialist who knows that that is what she is doing as it does to one who does not.

So we can take forward from this discussion that although the agent in a medical Newcomb case need not have precise knowledge of her beliefs and desires, she does know that (4.11) and (4.12) apply to herself. And this is something that she knows to be true whatever she ends up actually doing. Suppose that Susan, who knows that (4.11) and (4.12) are true, decides not to smoke. Then this tells her nothing about her prior beliefs and desires about this case that she didn't already know, *except* that she is in fact (at least as far as this case is concerned) sympathetic to evidentialism. Given that she knows (4.12), the only thing that smoking *could* tell her is that since evidentialism drove her to decline to smoke, she must already have taken smoking to be relatively bad news. But she already knew that because she already knew (4.11). She already knew that her desires and beliefs were such that smoking would be bad news.

4.3.3 *The conclusion*

Nor therefore does her actually declining to smoke tell her anything about whether she has the lesion. In particular it is no evidence *for her* that she *lacks* the lesion. The reason is that by (4.7), the absence of the lesion only encourages non-smoking in *Smoking Lesion* in so far as it has an impact on the desires and beliefs that are relevant to whether she smokes in that case. And by (4.8), she already knows enough about those desires and beliefs to make actual abstention evidentially irrelevant to *any* causal influence upon them.

The argument for that takes the following form. Turning the key in the ignition causes the engine to start, which in turn causes the car to move if the clutch is engaged and the handbrake is off. Then seeing the car move gives a bystander reason to think someone has turned the key: but not if he can already hear the engine running. More generally, the principle behind the argument is that an effect is evidentially irrelevant to any of its distal causes if we are already given an independent and fixed opinion about some intermediating proximal cause. In the jargon: an intermediary I screens off its effects E from causes C that only operate upon E through I.[16]

In this case the effect is abstention (motion of the car); its distal cause is the absence of the lesion (turning of the key); the intermediating proximal cause is smoking's being bad news (running of the engine). There is also a non-intermediating proximal cause, or causal condition, which Susan *may* learn about from abstention, for all that I have said here: this is an inclination towards EDT. But in *Smoking Lesion* this is irrelevant to the absence of the lesion. Similarly, seeing the car move *does* tell you that the clutch is engaged and the handbrake is off, even if you can hear the engine

---

[16] This follows from the causal or parental 'Markov condition' that (roughly) an event is evidentially independent of its non-effects conditional on its immediate causes (Bovens and Hartmann 2003: 68–9). Cartwright (1999: 7) presents an apparent counterexample to the principle. The version of *Prisoners' Dilemma* that I discuss at section 4.6 below, and perhaps also van Fraassen's version (as reported in Jeffrey 1983: 20; but see n. 11 above), can be seen as exploiting the structure of her counterexample to give a version of Newcomb's problem against which the Tickle Defence is unavailable.

running. But if you already know that the engine is running then the car's moving is still evidentially irrelevant to whether the key has been turned.

More generally, if Susan already knows enough about her beliefs and desires to be able to explain in terms of them why she would do such-and-such, then doing such-and-such tells her nothing either about them *or* about their causes. That she takes smoking to be bad news already tells her why she would abstain if she does. So if she *does* abstain then she gets no further confirmation of whatever caused her to take smoking to be bad news, so no further confirmation of the absence of the lesion beyond what (4.11) already gave her. It follows that when she is deliberating about whether to smoke, her credences must conform to:

(4.13) $Cr(S_1|O_1) = Cr(S_1)$.

Since O_2 is the only alternative to O_1 and S_2 is the only alternative to S_1, it follows from (4.13) that:

(4.14) $Cr(S_1|O_2) = Cr(S_1)$
(4.15) $Cr(S_2|O_1) = Cr(S_2)$
(4.16) $Cr(S_2|O_2) = Cr(S_2)$

And (4.13)–(4.16) entail the conclusion of the Tickle Defence, which is that whatever Susan does in *Smoking Lesion* is for her evidentially irrelevant to whether or not she has the lesion. It also follows from (2.18) and (4.13)–(4.16) that Evidential Decision Theory agrees with Causal Decision Theory in this case: it advocates smoking. So it follows, finally, that *Smoking Lesion* is not a Newcomb problem at all and constitutes no realistic case for causalism.

Three comments on this argument. (i) The move that I just made, from the premise that abstention is no sign of the absence of the lesion, to the conclusion that smoking is no sign of its presence, is entirely trivial on the Bayesian conception of evidence. But we can also reach that conclusion more intuitively. Since Susan already knows that she is facing a Newcomb problem, she knows that she prefers smoking with the lesion to abstention with it, and that she prefers smoking *without* the lesion to abstention without it. But knowing this, she knows that if she does smoke then these preferences would explain that by themselves. Her decision to smoke in this situation therefore carries no further information about any beliefs and desires underlying those preferences, and so none about the presence of the lesion. Smoking might carry that information in other situations, for instance in a situation where she knows that the lesion causes smoking but has no idea of its harmful side effects. But that would not be any sort of Newcomb problem.

(ii) You might be concerned that the argument is self-undermining. After all, once we have shown that Susan's credences conform to (4.13)–(4.16) we have also shown that smoking is not bad news for her after all. So how can she know – as I required in the argument for the transparency premise – that smoking *is* bad news for her?

The 'tickle' argument can be seen as a *reductio* of the idea that *Smoking Lesion* is a Newcomb problem. Given the description of Susan's case, we are entitled to assume that Susan does take smoking to be bad news. If the conclusion is that she *doesn't* take smoking to be bad news then this doesn't undermine the argument itself but only the assumption that the description of the case is coherent. In fact it is not. Nobody who knew even the little that Susan knows about herself could think (a) that smoking is a sign that the lesion is present, *and* (b) that this is because the lesion affects Susan's relevant desires and beliefs, *and* (c) that her desires and beliefs give this case the structure of a Newcomb problem.

(iii) The Tickle Defence presented here applies specifically to medical cases in which it is plausible that the undesirable event produces the symptomatic option via the agent's prior beliefs and desires, not by inclining the agent towards a particular decision rule or through some other mechanism. So it only shows that *these* cases fail to be realistic Newcomb problems. There are other apparently real-life Newcomb cases to which the argument does not apply in this form. But we can, and the next two sections do, address some of these other cases in a more or less ticklish spirit.

4.4 Economic cases

The first type of problem, here called *Expand or Contract*, arises from the 'rational expectations' theory of economic agents. According to that theory, agents make on average reasonably accurate predictions of the future values of economic variables, the error term being purely random and having an expectation of zero.[17]

In particular, suppose that the average public expectation of money supply is in general a good predictor of it. So for instance, if on some occasion the monetary authority (say, the Federal Reserve) increases the rate of growth of the money supply, consumers will have expected this. Similarly, if the authority reduces that rate or keeps it constant then consumers will have expected that. Let the agent be the monetary authority and let it be facing the choice between (O_1) increasing the rate of growth ('expansion') and (O_2) not increasing it ('contraction').

---

[17] Muth 1961. For references to empirical data, and a brief discussion of the subtleties involved in interpreting them, see Kydland and Prescott 1977: 473–91.

Table 4.2 *Expand or Contract?*

	S_1: expects contraction	S_2: expects expansion
O_1: contract	10	0
O_2: expand	11	1

According to a simple and well-known macroeconomic model, the effect of the monetary authority's decision will depend on whether or not the public anticipated it.[18] In particular: (a) an unanticipated expansion will cause a modest rise in inflation and a fall in unemployment; (b) an anticipated expansion will cause a sharp rise in inflation whilst unemployment remains constant; (c) an unanticipated contraction will cause a modest fall in inflation and a rise in unemployment; (d) an anticipated contraction will cause a sharp fall in inflation whilst unemployment remains constant. It is easy to imagine an authority whose trade-offs between inflation and unemployment imply the following ranking of these outcomes: (a) $\succ$ (d) $\succ$ (b) $\succ$ (c). We can therefore assign notional utilities to these outcomes as in Table 4.2.

This problem clearly has the dominance structure of a Newcomb problem in which contraction corresponds to one-boxing and expansion corresponds to two-boxing. It is also true that the authority's decision is causally irrelevant to the public's expectation of it. But it looks *evidentially* relevant to that expectation given *rational* expectations. That explains why some writers have seen this case, and other economic decisions that similarly involve rational expectations, as real-life versions of the Newcomb problem.[19]

But this appearance starts to dissolve when we consider the basis on which the public is supposed to form its expectation of monetary policy. The basis can only consist in evidence that is also available to the monetary authority itself. In general we may expect it to consist in facts about the past values of the relevant economic variables; about the past behaviour of the monetary authority and others like it in similar situations; about the authority's interests, beliefs and expectations (as revealed in policy documents, forecasts and other statistics); about its own policy rules; and, finally, about any explicit commitments that the authority has made.[20] Call

---

[18] For details see Frydman, O'Driscoll and Schotter 1982: 314–18.
[19] Frydman, O'Driscoll and Schotter 1982; Broome 1989.   [20] Kydland and Prescott 1977: 478.

the aggregate of this information E; call the proposition that this represents the totality of relevant and publicly available information P_E.

The monetary authority should take P_E to screen off any decision that it makes from the public expectation of that decision. After all, no cause of the actual decision can affect public expectation of it except via its impact on information that was publicly available *ex ante*. So the policymaker's credences should conform to:

(4.17) $Cr(S_1|O_1 \wedge P_E) = Cr(S_1|O_2 \wedge P_E) = Cr(S_1|P_E)$
(4.18) $Cr(S_2|O_1 \wedge P_E) = Cr(S_2|O_2 \wedge P_E) = Cr(S_2|P_E)$

And, since it is known to the policymaker just what evidence is available to the public, P_E forms part of the policymaker's evidence; hence:

(4.19) $Cr(S_1|P_E) = Cr(S_1)$
(4.20) $Cr(S_2|P_E) = Cr(S_2)$

It follows from (4.17)–(4.20) that $V(O_2) > V(O_1)$, i.e. that Evidential Decision Theory recommends the 'two-boxing' option of expansion. The reason for this is that if we are given P_E, as in this model we are, public expectations become *evidentially* independent of the monetary authority's actual decision, in which case *Expand or Contract* fails to be a genuine Newcomb problem. This argument is essentially a simplified version of the Tickle Defence in which public knowledge of P_E is a cause of public expectation that evidentially screens it off from the choice to expand or to contract.

Broome, who should take credit for introducing this case to philosophers, has stated – wrongly in view of the foregoing – that the discussion of government decisions in the face of rational expectations has relevance to Newcomb's problem (in which case, as he says, most economists are two-boxers[21]). But *Expand or Contract* has an economic interest that is quite independent of any resemblance that it bears to the Newcomb problem. It raises a problem for the *credibility* of policy announcements.

Suppose that, in order to get the best outcome (i.e. $O_2 \wedge S_1$), the monetary authority states in advance that it is determined to contract. If the public believes this announcement, then when the moment of truth arrives it is in the authority's interests to *renege* on that commitment (assuming that this is a one-shot decision), for at that point it effectively faces a choice between $O_1 \wedge S_1$ and $O_2 \wedge S_1$, the latter being unambiguously preferable and so true. But everybody knew this in advance. So there was no reason in advance to believe the announcement. This is the problem of credibility.

---

[21] Broome 1989: 222.

The problem does of course depend on the public's, or markets', having rational expectations in the sense that these actors are able to calculate in advance what it would be in the monetary authority's interest to do, and so to adjust their own actions in the expectation that this *is* what the authority will do. But it does not follow that at the point of decision the authority should think that *whatever* it then does, everyone expected it to do just *that*. On the contrary, if the authority announces a contraction then it will know at the point of decision that everyone will expect *expansion*, whatever it in fact then does. In that case, EDT and CDT *agree* on expansion. So the case cannot have the relevance to decision theory that Newcomb's problem is supposed to have.

As recent Eurozone crises illustrate, credibility is a genuine problem for governments, central banks and other large economic actors. But it has not got anything to do with Newcomb's problem.

4.5   Psychological cases

Probably the best-known psychological Newcomb problem is a fictitious example in which a businessman, who hopes to have scored well for ruthlessness in a test that he has just taken, must now choose whether to fire a long-standing and loyal employee whose current underperformance is probably temporary.[22]

A better (because real) example is the experimental version of Newcomb's problem that Shafir and Tversky reported in their 1992 paper. In stage 1 of this experiment, the authors presented each of forty Princeton undergraduates with forty computerized games, of which six were versions of *Prisoners' Dilemma* and the other thirty-four were of other types whose details don't matter. (For more on *Prisoners' Dilemma* see 4.6 below.) The rewards in each game were 'points' corresponding to final monetary rewards. In each game a subject chose an option by pressing a button on the computer.

At stage 2 subjects received the following text:

> ***P-Newcomb***: You now have one more chance to collect additional points. A
> program developed recently at MIT was applied during this entire session to

---

[22] Gibbard and Harper 1978: 354. See also their 'King Solomon' case (ibid.: 353–4). The discussion in this section applies to both cases. It also applies to Cartwright's (1979: 419–20) example involving life insurance, at least in one plausible way of filling this out. And it applies to the cases with which McKay (2007: 397) objects to Price's (1991) version of the Tickle Defence, in which subconscious motives produce inclinations to act one way or another.

analyse the pattern of your preferences. Based on that analysis, the program has predicted your preference in this final problem.

[Box A: 20 points]   [Box B: ?]

Consider the two boxes above. Box A contains twenty points for sure. Box B may or may not contain 250 points. Your options are to:

(1) Choose both boxes (and collect the points that are in both).
(2) Choose Box B only (and collect only the points that are in Box B).

If the program predicted, based on observation of your previous preferences, that you will take both boxes, then it left Box B empty. On the other hand, if it predicted that you will take only Box B, then it put 250 points in that box. (So far, the program has been remarkably successful: 92 per cent of the participants who chose only Box B found 250 points in it, as opposed to 17 per cent of those who chose both boxes.)

To insure that the program does not alter its guess after you have indicated your preference, please indicate to the person in charge whether you prefer both boxes or Box B only. After you indicate your preference, press any key to discover your allocation of points.[23]

In this problem, the statistics concerning the predictor's success are easily believable. Interviewed after the event, no subject showed any suspicion that the experimenters were subtly cheating. And if we exclude cheating then there is no possibility that the prediction is *causally* dependent on the choice.

The Tickle Defence as applied to *Smoking Lesion* and *Expand or Contract* seems not to get any traction in this case. The subject is simply ignorant of what algorithm the computer program uses to predict his choice; also ignorant, therefore, of both the nature and the mode of operation upon himself of the common cause of both that prediction and his eventual choice. In particular, there is no special reason to think that whatever psychological state of his the test detected was something that encourages two-boxing in *P-Newcomb* via its effects on the agent's *beliefs and desires*. It is more plausible that the test identifies not the agent's beliefs and desires but rather his inclination towards a decision rule. (In fact, an algorithm that predicts that an agent will one-box given at least two 'evidentialist' (i.e. co-operative) responses to *Prisoners' Dilemma* would, the authors note, have had a real-life success rate of about 70 per cent.[24]) And whereas the agent probably does know something about the beliefs and desires that he brings to this case, he might *not* have any special knowledge

[23] Shafir and Tversky 1992: 461–2.   [24] Shafir and Tversky 1992: 462.

that he inclines to this or that decision rule. Certainly we cannot assume this on the sole grounds that the agent knows that he is facing a New-comb problem. Still, something *like* the Tickle Defence does apply to *P-Newcomb*.

A deliberating agent is by definition one who hasn't yet decided what he will do. But he does, whilst in that mode, have a reasonably good idea of the state of his deliberation: how strongly he is currently inclined in one or another direction. For instance, the deliberator in *P-Newcomb* knows either that he is strongly inclined towards taking both boxes, or that he is strongly inclined towards taking just Box B, or that he is not strongly inclined in either direction. He might even be able to measure the strength of any strong inclination in either direction e.g. if he knows how much compensation he would demand to take the other option right now. This knowledge that the deliberator has of his own deliberative state is perfectly commonplace and is not confined to *P-Newcomb*; probably if you interrupted anyone in the course of deliberation he would be able to tell you how he was then inclined to act, at least in this relatively coarse-grained way.

An analogy of this sort of knowledge arises in the purely theoretical case where one is deciding what to think on some question. If you are trying to work out the solution to some chess puzzle, your opinion as to the best next move for white might evolve from being in favour of one move to being undecided to being in favour of another move, before being undecided again and finally settling on the first move. But at any point in this process you would be able to say what state your evolving opinion had reached, at least in a coarse way ('I strongly suspect that this is the best move'; 'I have no idea which of these three moves is best', etc.[25])

In the practical case and in this purely doxastic one, what puts the agent in a position to know is not some Cartesian inner vision. It is that in both cases the subject answers (a) the 'internally' directed question 'What am I now inclined to do/think?' *by* answering (b) the 'externally' directed question 'What *should* I do/think?' As Edgley puts it, (a) is, from the first-personal perspective, *transparent to* (b).[26] Being able to answer (b) *sincerely* already puts you in a position to answer (a) *truly*. And answering

---

[25] This example relaxes the idealization imposed at section 1.1 (b2) that the agent should be ideally rational. Ideally rational agents know the answers to all chess problems in advance. Since the only role it plays in the argument is that of an analogy, this is harmless.

[26] Edgley 1969: 90. See also Hampshire 1975: 70; and Evans 1982: 225. The passages from Edgley and Evans appear, and are discussed, in Moran 2001: section 2.6. Hampshire 1975: Chapter 3 is a careful treatment of the practical case.

(b) sincerely requires no Cartesian superpowers but only some kind of responsiveness to the factors that speak for and against this or that action or conclusion. This kind of transparency therefore constitutes a second reason to accept the transparency premise of the Tickle Defence: see section 4.3.2 above.

I should note that the practically deliberating agent's knowledge of his current inclinations is not in itself, although it might ordinarily ground, knowledge of any purely future-tense proposition concerning what he *will in fact* do. (Compare: knowledge of his current doxastic inclinations is not in itself knowledge about what he will in fact end up believing.) It may be true that the deliberating agent is often in some special position to know the answer to that genuinely future-tense question. But the transparency relations that I am here discussing do not establish this. Nor does it matter at this point in the discussion. It *does* matter at section 8.2, where it gets further treatment.

Now the instructions in *P-Newcomb* are deliberately vague on the actual means by which the predictive program works. But suppose that the subject believes the following hypothesis:

(4.21) **Screening by inclination (SI):** The psychological test predicts how I will act by identifying what causes my initial *inclinations*.

If the subject is certain of **SI** then he should take his current inclinations to make his ultimate action evidentially irrelevant to the result of the test, just as in *Smoking Lesion* Susan may take her current feelings about cigarettes to make her ultimate decision whether or not to smoke to be evidentially irrelevant to whether or not she has the lesion, and just as in *Expand or Contract* the monetary authority takes publicly available evidence of its decision to make its actual decision evidentially irrelevant to the public expectation of it. And just as in those cases – the details are the same so I won't go into them – the subject's knowledge of his current inclinations means that Evidential Decision Theory agrees with CDT that the subject should take both boxes. *P-Newcomb* is then *not* a genuine Newcomb problem.

This version of the Tickle Defence will work if the subject is *completely certain* of the hypothesis **SI**. But it will also work if the subject is *confident enough* of **SI**. If the utility of points is linear and the program correctly predicts proportions n_1 and n_2 of one-boxers and two-boxers respectively, and if the ratio of the potential amount in Box B to the known amount in Box A is R, then EDT endorses taking both boxes if $Cr(\mathbf{SI}) \geq 1 - 1 / R(n_1 + n_2 - 1)$. This is compatible with uncertainty about

SI; in fact on the particular figures that Shafir and Tversky give it suffices that $Cr\,(\text{SI}) \geq 0.9.$[27]

Whether or not the Tickle Defence works against *P-Newcomb* therefore turns on whether the subject realistically has a high confidence in SI. Horwich thinks we should be certain of it in *all* realistic cases of psychologically based (and physiologically based) prediction of human action:

> [T]he causal relationships between a person's antecedent physiological and psychological states and his subsequent action are often extraordinarily complex, subtle and difficult to recognize. Those very few causal generalizations that are manifest to ordinary observation or current psychology are bound to be crude and blatant. They certainly do not include actions resulting from relatively complex bouts of deliberation. Rather, they are restricted to simple cases where a physiological or psychological condition tends to cause an action (in a given sort of situation) by tending to cause either certain characteristic desires, or certain characteristic beliefs, or conformity to a particular decision rule, in the light of which it becomes fairly obvious what should be done. Consequently, if an agent knows that a condition CA has tended to result in action A, he can be sure that A was the result of a simple deliberation process whose salient features will be relatively easy to discern.[28]

[27] Write O_1 and O_2 for one-boxing and two-boxing respectively. Write P_1 and P_2 for the propositions that the program predicted one-boxing and two-boxing respectively. EDT endorses two-boxing if and only if:

   (i) $Cr\,(P_1|O_2) - Cr\,(P_1|O_1) \geq -1/R.$ But:
   (ii) $Cr\,(P_1|O_2) = Cr\,(P_1|O_2 \wedge \text{SI})\,Cr\,(\text{SI}|O_2) + Cr\,(P_1|O_2 \wedge \neg\text{SI})\,Cr\,(\neg\text{SI}|O_2);$ and
   (iii) $Cr\,(P_1|O_1) = Cr\,(P_1|O_1 \wedge \text{SI})\,Cr\,(\text{SI}|O_1) + Cr\,(P_1|O_1 \wedge \neg\text{SI})\,Cr\,(\neg\text{SI}|O_1).$

Since SI makes the agent's choice evidentially irrelevant to the prediction we have $Cr\,(P_1|O_2 \wedge \text{SI}) = Cr\,(P_1|O_2 \wedge \text{SI}).$ And since the agent's actual choice is evidentially irrelevant to the theoretical question whether SI is true, we have $Cr\,(\text{SI}|O_1) = Cr\,(\text{SI}|O_2) = Cr\,(\text{SI}).$ So substituting (ii) and (iii) into (i) gives the following sufficient condition for (i):

   (iv) $Cr\,(\neg\text{SI})\,(Cr\,(P_1|O_2 \wedge \neg\text{SI}) - Cr\,(P_1|O_1 \wedge \neg\text{SI})) \geq -1/R.$

Assuming that $\neg\text{SI}$ makes the agent's actions evidentially relevant to the prediction in accordance with the known statistical record, because there is in that case no screening-off, we have:

   (v) $Cr\,(P_1|O_1 \wedge \neg\text{SI}) = n_1$
   (vi) $Cr\,(P_2|O_2 \wedge \neg\text{SI}) = n_2.$

Substituting (v) and (vi) into (iv) gives the following sufficient condition for (iv):

   (vii) $Cr\,(\neg\text{SI})\,(1 - n_2 - n_1) \geq -1/R.$

If the program is even minimally reliable then $n_1 + n_2 - 1 > 0$ so it suffices for (vii) that:

   (viii) $1\,/\,R\,(n_1 + n_2 -1) \geq Cr\,(\neg\text{SI}).$

Hence it suffices for (i) that $Cr\,(\text{SI}) \geq 1 - 1\,/\,R\,(n_1 + n_2 -1).$ Shafir and Tversky's figures imply that $R = 12.5$, $n_1 = 0.92$ and $n_2 = 0.83.$ So EDT recommends two-boxing if $Cr\,(\text{SI}) \geq 0.9.$ Of course we can drive this quantity arbitrarily high by simply assuming R to increase without limit. But as R increases intuition tends more and more to favour one-boxing in any case (Nozick 1993: 44).

[28] Horwich 1987: 182.

If Horwich is right then it is reasonable to be very confident of **SI**. The subject will therefore quite reasonably take his own options to be evidentially irrelevant to the prediction in *P-Newcomb* once given his present inclinations. More specifically, suppose that I am deliberating over whether to take only box B or to take both boxes. I find myself strongly inclined to take two. Then I should already be confident that the program has predicted this. What I then go on to do is not in itself any further evidence of that prediction, at least not to me now. The same is true if I find that I am strongly inclined to take just one box. Finally, if I find myself undecided, then that too is decisively revealing, since it means that whatever straightforward deliberative process the program predicted has simply not occurred. And so again I have no reason to think that its ultimate upshot is any sign of the contents of Box B.

I do not see that it can be settled a priori whether psychological prediction will ever surpass the relatively primitive stage that Horwich describes. Still, in the absence of evidence that it has, it is reasonable to be confident of **SI** and so to deny that any realistic version of *P-Newcomb* is a genuine Newcomb problem. Certainly nothing in the problem as described by its authors gives us any reason to think that the predictive mechanism is both realistic enough and sophisticated enough to block this modified version of the Tickle Defence.

A final comment on the Tickle Defence in general. Its basic strategy is to find some proposition that (a) is known to the agent and (b) screens off what the agent does from the allegedly correlated event. That the defence was conducted from the agent's point of view was of course a prerequisite on its working at all. In the present subjective framework, all judgements of practical rationality are made against the background of whatever evidential and causal relations are visible from that perspective. But the connection between the Tickle Defence and the 'agent's perspective' does not run any deeper than that.

First, although the agent has knowledge of the screening-off proposition, that knowledge needn't be *exclusive* to the agent. You don't even have to be *an* agent, let alone *that* agent, to have it. Spectators could also come to know the screening-off proposition. And if they do, then the agent's act has no evidential bearing on the prior event for them either. This is perhaps most obvious in *Expand or Contract*. The key proposition P_E, concerning what information is available to the public, may be itself available to the public – whether it is makes no difference to the argument. But it is also true in the medical and psychological cases. Any observer could learn of

Table 4.3 *Prisoners' Dilemma*

	B_1: Bob leaves the \$1K	B_2: Bob takes the \$1K
A_1: Alice leaves the \$1K	(M, M)	(0, M + K)
A_2: Alice takes the \$1K	(M + K, 0)	(K, K)

Susan's feelings about smoking, or the inclinations of one of Shafir and Tversky's subjects, by simply asking.

Second, nothing about being an agent *guarantees* the availability of a screening-off proposition. That it is possible to find such propositions in these cases is a contingent feature of the 'predictive' mechanisms involved. There *might* have been, and there might one day be, psychological tests that predict the upshot of deliberation independently of whatever gross inclinations precede it. And so the fact that *these* predictive mechanisms are, if realistic, subject to the Tickle Defence does not rule out the existence of predictive mechanisms against which there is *no* Tickle Defence. I turn to one such now.

4.6 *Prisoners' Dilemma*

Alice and Bob are strapped to chairs facing in opposite directions. In front of each of them is 1,000 dollars. Each must now choose between taking it and leaving it. Each of them will receive 1 million dollars if and only if the other player *doesn't* take the 1,000 dollars. Writing the outcomes in the form (Alice's pay-off, Bob's pay-off) we have Table 4.3.

Everyone will recognize this situation as *Prisoners' Dilemma*. Although the situation is unrealistic it is certainly feasible. And in fact *Prisoners' Dilemma*-type situations frequently arise in real life. For instance, replace Alice and Bob with two competing firms; replace the decision whether to take the thousand dollars with the decision whether to launch a costly advertising campaign.

4.6.1 *Is it a Newcomb problem?*

The present interest of the case is that on certain additional assumptions, each player is facing a version of the Newcomb problem.[29] Focus on Alice.

---

[29] Lewis 1979b.

For her to be facing a Newcomb problem three things must be true: (a) one of her options dominates the other with respect to Bob's choice; (b) what Bob does is causally independent of what she does; (c) EDT recommends the dominated option to Alice.[30] Condition (a) is straightforwardly true: on either hypothesis about what Bob does, taking the 1,000 dollars leaves her that much better off than leaving it. So is (b): since Alice and Bob choose simultaneously and cannot communicate, Alice's choice cannot affect Bob's. It follows straightforwardly from (a) and (b) that CDT endorses the dominant option. It advises Alice to take the 1,000 dollars.

Condition (c) is equivalent to the claim that Alice takes her choice to be (sufficiently) evidentially relevant to Bob's, in particular that her taking the 1,000 dollars is a sign that Bob will do the same. This condition is not explicit or implicit in standard presentations of *Prisoners' Dilemma*,[31] which is hardly surprising given its usual game-theoretic, not decision-theoretic, context. And why should (c) be realistic? After all, I don't normally take my actions to be evidentially relevant to anyone else's. If I choose to bet aggressively in this poker hand, I won't take this as even the slightest reason to expect similar aggression from my opponents. That is why I expect it to work.

Lewis claims that Alice will take her own choice as at least slightly indicative of Bob's if she considers Bob to be a perhaps very imperfect and perhaps very unreliable replica of herself. He writes:

> The most readily available sort of replica of Alice is simply another person placed in her predicament. For instance: Bob, her fellow prisoner. Most likely he is not an exact replica of her, and his choice is not a very reliable predictive process for hers. Still, he might well be reliable enough (in her estimation) for a Newcomb problem.[32]

Lewis is right that *if* Alice takes Bob to be even a slightly reliable simulacrum then we have a genuine Newcomb problem, i.e. in which (c) is true. In fact

---

[30] Lewis's 1979 paper on *Prisoners' Dilemma* does *not* require that condition (c) be satisfied for it to count as a version of Newcomb's problem. He takes (a) and (b) to suffice (Lewis 1979b: 303 n. 6; but he takes a different view in his 1983: 533). But in that looser sense, a decision situation can qualify as a Newcomb problem even if EDT and CDT agree in recommending the dominant option. *These* 'Newcomb problems' are of no interest in the present context, since they have no bearing on the dispute between EDT and CDT. Here and elsewhere, what *I* mean by 'a version of Newcomb's problem' is what Lewis would mean in his 1979 usage by 'a Newcomb problem over which EDT and CDT disagree'.

[31] Kreps 1990: 28–9; Osborne and Rubinstein 1994: 16.

[32] Lewis 1979b: 303, except that I have 'Alice' and 'Bob' for Lewis's 'me' and 'you' etc.

if Alice has linear utility for dollars, and thinks that Bob is equally likely to do what she does, whatever she does, then it suffices for (c) that:

$$(4.22) \quad Cr_{\text{ALICE}} (B_1 \mid A_1) = Cr_{\text{ALICE}} (B_2 \mid A_2) > 0.5005.^{33}$$

But why should Alice's credences follow (4.22) in any realistic case? Why would it be more reasonable for her to think it more likely that Bob is an imperfect simulacrum of her than that he is an imperfect 'anti-simulacrum'? One possible reason would be that Alice and Bob are antecedently alike in terms of character, habits, upbringing, etc. Thus some writers suppose that they are twins.[34]

But the inference from this premise to the reasonableness of (4.22) seems to face a tickle-style objection along the lines of section 4.5. Their *being* alike would presumably cause their now *acting* alike by causing them to be similarly inclined. But since Alice knows her own inclinations whilst deliberating about what to do, the news that she does in fact take the 1,000 dollars is for her no *additional* evidence that Bob is doing the same. So their being even *very* alike *ex ante* is no reason to think that, whilst she is deliberating, Alice's credences will reasonably conform to (4.22).

So there is no obvious reason to treat *Prisoners' Dilemma* as a realistic Newcomb problem. *Prisoners' Dilemma* is *realistic* all right. But that just means that we frequently encounter two-person strategic interactions analogous to Table 4.3 that from both players' points of view satisfy analogues of (a) and (b). But these are *not* Newcomb problems because from neither point of view do they satisfy (c), since neither party reasonably takes its own action as evidentially relevant to the other's. Or so it seems.

4.6.2   Consensus effect

It should give us pause that educated and intelligent individuals frequently *do* take their own actions (and beliefs) to be predictive of the actions and beliefs of causally isolated others in the same situation. This is what psychologists call the *false consensus effect*: 'people who engage in a given behaviour will hold that behaviour to be more common than it is estimated to be by the people who engage in the alternative behaviour'.[35]

There is plenty of evidence that this effect is widespread. And it persists even amongst individuals who know that their own choices are *causally* irrelevant to anyone else's. For instance, in the original study of false consensus:[36]

---

[33] $V(A_1) > V(A_2)$ if $MCr(B_1|A_1) > K + MCr(B_1|A_2) = K + M - MCr(B_2|A_2)$. So if $Cr(B_1|A_1) = Cr(B_2|A_2)$ then $V(A_1) > V(A_2)$ if $2MCr(B_1|A_1) > M + K$ i.e. if $Cr(B_1|A_1) > 0.5 + K/2M$.

[34] E.g. Nozick 1969: 223–4.   [35] Mullen et al. 1985: 262.   [36] Ross, Greene and House 1977.

Stanford students were asked to engage in a number of activities, one of which was to walk around the Stanford campus with a big sign reading 'Repent!' Students either agreed to engage in this activity or refused and then were asked to estimate the proportion of Stanford students who would agree. The students who agreed made an average estimate that 63.5% of Stanford students would agree, whereas, among those who refused, the average estimate was 23.3%.[37]

Of course the purely descriptive question whether people do in fact take their actions to be diagnostic of others' is independent of the question whether this is rational. If it is in fact an error, arising for instance from an erroneous, superstitious and perhaps only half-conscious belief that one's own actions *influence* those of others, then it has no further interest. The aim of this chapter was to find a version of Newcomb's problem that does not require the agent to have fantastic or science-fictional beliefs about his situation. That a fantasy is common doesn't make it any less a fantasy.

But somewhat confusingly in light of its name, false consensus effect *need* not involve any irrationality at all. Notwithstanding the fact that the sample size is just one, it may be prima facie rational to alter your beliefs about others on the basis of observation of yourself (as when Jonas Salk tested polio vaccine on himself) – and why shouldn't this include observations of what you *do*? Genuinely biased extrapolations from oneself are what psychologists call the *truly* false consensus effect.[38] But there is nothing inevitably irrational in revising your beliefs about what Bob will do given what Alice does – even if *you* are Alice.[39]

For instance, suppose that Alice thinks that what causes her and Bob to decide to take the 1,000 dollars is some type of chancy mechanism that they have in common. Of course this is a crude simplification. But it is hardly implausible that *some* common springs of action exist, and that is enough to motivate some such line of thought.

Alice doesn't know with what chance the mechanism *does* make her choose to take the 1,000 dollars. But noticing that she has done so is surely a good reason to raise her estimate of it. For instance, if she starts out with a uniform distribution over her and Bob's common chance of taking the 1,000 dollars, her actually taking the 1,000 dollars should raise her confidence that Bob will do the same from 0.5 to 0.67.[40] Such a

---

[37] Dawes 1990: 180. Wolfson 2000 documents a similar effect in connection with estimates of the prevalence of drug use amongst university students.

[38] Krueger and Zeiger 1993.   [39] Dawes 1990: 182.

[40] Suppose that c is the true common chance at the outset that Alice / Bob takes the 1,000 dollars. Then to say that Alice has a uniform distribution over this chance is to say that her $Cr\,(c \leq x) =_{\text{def.}}$ $F\,(x) = x$ for $0 \leq x \leq 1$. Now $Cr\,(A_2|c \leq x) = x^{-1} \int_{0 \leq t \leq x} t\,dF(t) = x/2$. So $Cr\,(c \leq x|A_2) =_{\text{def.}}$ $F_{A_2}\,(x) = x^2$. Finally, $Cr\,(B_2|A_2) = \int_{0 \leq t \leq 1} t\,dF_{A_2}(t) = 2/3$.

background belief is hardly fanciful. If anything it evinces the hard-headed detachment with which we should expect a scientifically minded person to think about herself as well as about others. That *Prisoners' Dilemma* can appear from *that* perspective to be a Newcomb problem speaks only in favour of its realism.

But won't the Tickle Defence apply here too? If, as you might think plausible, the mechanism influences Alice by operating on her *inclination* to take the 1,000 dollars, and if this is something for which she already has pretty robust evidence, shouldn't her actual decision to take the 1,000 dollars be evidentially irrelevant to what Bob does? Yes it should, *if* that is how Alice takes the mechanism to operate. But what if, instead, she thinks that what she and Bob share is a chancy mechanism of *deliberation*? In that case, neither knowledge of her own beliefs and desires at the outset of reasoning, nor knowledge of her inclinations at any intermediate stage of it, suffices to screen off her action from Bob's. For that, she would also need to know in which option these beliefs and desires eventuate. And she can only find out about that *by* deciding what to do. Prior to the decision, she must take either decision to be diagnostic of her part-ner's, because indicative of a common reasoning process that concludes with it.

An analogy: suppose that a factory manufactures biased chancy coins to the same specification. Let A and B be such coins, each being tossed once. The *outcome* of the A-toss, and not just the process leading up to this, will carry news about the outcome of the B-toss. It may be that no observation, however accurate, of the position, angular momentum etc. of coin A at any particular moment *before* it lands will screen off this evidential bearing of one outcome on the other. Nothing changes if the 'coins' are neuronal and their 'tosses' occur within people's heads.

So belief in common deliberative mechanisms is sufficient to constrain Alice's beliefs in such a way that *Prisoners' Dilemma* meets condition (c), since it makes Alice's actual choice seem to her to be a sign of Bob's choice even given her present inclinations. But it isn't necessary. Another quite common way of thinking about other minds has just the same effect.

I mean reasoning about others by *simulating* them, i.e. learning about what another will do in a given decision situation by reasoning through that situation as if *you* were in it. When the subject and the object of this reasoning are in *different* situations the simulation will normally be 'off-line'. That is, it will not eventuate in action on the part of the subject. If I am wondering whether someone relevantly like me will choose the cream bun or the salad, I might hypothetically deliberate about what *I* would choose; but the point of its being hypothetical is that *I* would not then reach for the cream bun on which I had hypothetically settled. There might *be* no cream buns within *my* reach.

But if Alice and Bob face the *same* situation then Alice's simulation of Bob may *not* be off-line. Instead, the very same on-line deliberative process that reveals to her the overall trajectory of her *own* practical reasoning can also and simultaneously reveal that of Bob's. As Morton writes, 'One's understanding of what the other person may do comes from the fact that one is running through the thinking that results in their actions; if it results in one's own actions as well nothing is essentially changed.'[41]

So if Alice *is* such a simulator then what she actually does, and not only what she presently thinks and wants, will be a sign for her of what Bob actually does, because she thinks that her future deliberation will be a model of his. In that case EDT will advise her not to take her 1,000 dollars, even though (a) whether Bob does, or doesn't, take *his* 1,000 dollars, she is in either case better off taking her 1,000 dollars; and (b) her choice doesn't *make* any difference to whether Bob takes his 1,000 dollars.

Nothing in this argument relies on its *actually* being the case that people take either approach to other minds, i.e. belief in common deliberative mechanisms or 'on-line' simulation. What matters is that doing so involves nothing far-fetched, nothing like the fantasy in which the standard Newcomb agent is supposed to have complete confidence.

If that is right, then although the Tickle Defence suffices to undermine the alleged realism of some versions of the Newcomb problem, it is *not* an a priori defence against the possibility of such realistic Newcomb problems as these. Return to the original *Prisoners' Dilemma*. Even whilst deliberating, Alice can reasonably enough think that some overall evolution of her deliberation, including its upshot in action, is diagnostic of Bob's thinking and then doing the same. If she *does* think this then *Prisoners' Dilemma* is a Newcomb problem against which no Tickle Defence is available.[42]

---

[41] Morton 2002: 121.

[42] J. L. Bermudez argues that *Prisoners' Dilemma* is not a Newcomb problem on the grounds that if it is, then Alice should be close to certain that she and Bob will do the same thing, and hence that the outcome almost certainly corresponds to the top-left or the bottom-right cell in the body of Table 4.3. 'In other words, she is committed to thinking that she is in a completely different decision problem – in particular, that she is in a decision problem that most certainly is not a *Prisoners' Dilemma*, because it lacks precisely the outcome scenarios that make *Prisoners' Dilemma* so puzzling' (Bermudez 2013: 428). But it is simply not true that in a genuine Newcomb problem Alice must be close to certain that she and Bob will do the same. All that is required is that Alice

4.6.3   Simulation and co-operation

So there is at least one realistic version of the Newcomb problem over
which EDT and CDT clash. EDT recommends 'one-boxing' in it, i.e.
that Alice decline her 1,000 dollars. CDT recommends 'two-boxing', i.e.
that she take it. In terms that are more usual when discussing *Prisoners'
Dilemma*: EDT recommends that Alice co-operate and CDT recommends
that she defect.

That seems to make the case a strong argument for preferring CDT to
EDT. After all, it is received wisdom in game theory that defection is the
only rational strategy in *Prisoners' Dilemma*. That is what accounts for its
apparently paradoxical consequence that two rational prisoners will each
do worse than two irrational prisoners. So this seems to be a real-life case in
which the evidential strategy does worse precisely because it attends to the
diagnostic bearing of Alice's action on Bob's whilst disregarding its causal
irrelevance.

The main response to this point will be in Chapter 7, where I discuss
the standard Newcomb problem. Everything that I say from section 7.3
onwards in defence of one-boxing will apply equally to co-operation in this
version of *Prisoners' Dilemma*. At least it will if I have been right to argue
that it really *is* a Newcomb problem – if I am wrong about that, then the
case presents no threat to EDT.

But three comments are appropriate here. First, the obvious reason for
Alice to prefer defection (taking the 1,000 dollars) in *Prisoners' Dilemma*
is that it dominates co-operation. Whatever Bob does, Alice is better off
defecting than co-operating. But in game theory just as in decision theory,
the principle that you should take the dominant strategy has its limits.
Consider a new game, **Extended Dilemma**, in which both players' options
and Alice's pay-offs are as before. The differences are that (i) Bob knows
Alice's move before he makes his own; (ii) Bob forfeits all the money
if he makes a *different* choice from Alice. The joint pay-offs are as in
Table 4.4.

In this game Alice should reason as follows. 'Suppose I leave the 1,000
dollars. Then Bob will know this, and so will leave it himself, so I'll end
up with 1 million dollars. On the other hand suppose I take the 1,000
dollars. Then Bob will know this and will take it himself, and so I'll end

---

takes her own action to have *some* positive evidential bearing on the proposition that Bob will do
the same thing. In particular, and with linear utility for money and pay-offs as described in Table
4.3, it suffices as (4.22) says that $Cr\,(B_1|A_1) = Cr\,(B_2|A_2) > 0.5005$. This is consistent with any
credence short of 0.4995 that the top-right or bottom-left cell will be realized and hence with Alice's
being *entirely* certain that she and Bob face a *Prisoners' Dilemma*.

Table 4.4 *Extended Dilemma*

	B_1: Bob leaves the $1K	B_2: Bob takes the $1K
A_1: Alice leaves the $1K	(M, M)	(0, 0)
A_2: Alice takes the $1K	(M + K, 0)	(K, K)

up with 1,000 dollars. So I should leave the 1,000 dollars.' But from Alice's perspective, taking her 1,000 dollars is still the dominant option. This shows that the dominance argument can't apply *everywhere*.

So how do we distinguish the cases where it applies from the cases where it doesn't? Here we face the same problem as arose in connection with dominance reasoning in the Savage framework (see the discussion at section 1.6). We can require (i) that Alice should apply the dominance principle whenever Bob's action is *causally* independent of hers (according to her information). Or we can require (ii) that she should only apply it in those cases where her action is *evidentially* irrelevant to Bob's. Game theory itself does not adjudicate on this matter: to do so we must already have settled the issue between EDT and CDT. So the dominance principle that is standardly applied to *Prisoners' Dilemma* does *not* vindicate 'two-boxing' *in the non-standard version* of that problem over which EDT and CDT disagree.

If anything, *Extended Dilemma* seems to favour EDT. Alice's reasoning for not taking the 1,000 dollars looks plausible, but nothing in it relied on her choice being *causally* relevant to Bob's. It *might* be that in *Extended Dilemma* Bob learns what Alice does by seeing it. And in that case his choice *is* causally dependent on hers. But for all that I said about *Extended Dilemma*, Bob *might* be a Laplacean predictor who calculates Alice's choice in advance. Even then dominance reasoning looks intuitively wrong. Without pressing that point against CDT, we can at least conclude that there is no straightforward and non-question-begging argument against co-operation in the special version of *Prisoners' Dilemma* over which EDT and CDT disagree. This is entirely consistent with the fact, on which evidentialists can agree with everyone else, that defection is uniquely rational for both players in the standard version of the problem.

The second point is that the right arrangement of examples can encourage the two-boxing intuition to go away.[43] (i) Suppose first that you are

---

[43] Cases (i)–(iv) are variants of ones discussed in Leslie 1991.

facing a mirror and that in front of you, and so also 'in front' of your mirror image, are two buttons: one red, one blue. Pressing the red button causes 1,000 dollars to be wired to your bank account whatever happens. But if, and only if, your *mirror image* 'presses' the *blue* button, 1 million dollars will be wired to your account. What do you do?

(ii) You start to become a little suspicious about the mirror when you notice an unobtrusive wire running from its edge into a wall socket. It dawns on you that this is no mirror! It is a screen connected to Prof. Laplace's supercomputer, capable of predicting the future down to the finest detail. What you see *looks* for all the world like your mirror image. And you are practically certain that it will continue to behave just like your mirror image. But it is actually a fantastically detailed *simulation* of you. Now what do you do?

(iii) You recall that although Laplace's computer always gets it right, occasional irregularities in the current mean that its signal sometimes gets scrambled. This is very rare. But it means that if you were to play the game once every minute then perhaps once every ten years your simulation will 'press' a different button from the one that you press. Now what do you do?

(iv) You recall that the electricity workers are on strike and disruptions are very likely. This means that the signal will get scrambled very much more often, perhaps on one play out of every hundred. Now what do you do?

Everyone agrees that in case (i) you should press the blue button, since that would cause the mirror image to do the same. The step from case (i) to case (ii) is the step from a causally dependent mirror image to a causally independent simulation. At this point CDT says that that you should press the *red* button. Although you are just as certain as before that the image will behave *just like* a mirror image, it is (unlike a real mirror image) causally independent of you. But EDT still recommends pressing the blue button. If the simulation behaves just like a mirror image, then you may as well treat it like one. And I think that there is some intuitive pressure in this direction too.

It is reasonable to ask the causalist at this point just what it is *about* causal dependence that he thinks makes for a decision-theoretic difference between case (i) and case (ii). Why should the absence of this metaphysical connection, whatever it is supposed to be, make for any *practical* difference between a mirror and something that behaves exactly like a mirror? I will not press this difficulty here because case (ii) is unrealistic; it will arise again at section 6.4 in connection with the more feasible quantum scenarios that

that chapter discusses. For the moment the point is only that defection in a simulatory *Prisoners' Dilemma*, which is equivalent to pressing the red button in (ii), is not so robustly intuitive when the simulation is described like this.

The movement from case (ii) to cases (iii) and (iv) is itself an attempt to simulate greater realism, at least with respect to the accuracy of the 'mirror image'. There is no good reason to think that incrementally reducing the reliability of the simulation should make any difference to whether you should press the red or the blue button, at least not as long as it remains above a reasonable threshold (say, 51 per cent). In fact above that threshold this is something on which CDT and EDT agree. Nor should it make any difference if at this point we replace the simulatory mechanism with an equally reliable but more mundane one, so that instead of an image on a computer screen we have another person, or persons, who are at least somewhat like you. And now we have arrived at a realistic simulatory version of *Prisoners' Dilemma*, which is equally a realistic version of the Newcomb problem. And in contrast with the medical versions discussed at section 4.2, it is not so clear that 'two-boxing' *is* in this case the best strategy.

The third point is that there is also more direct evidence of a widespread 'one-boxing' or co-operative intuition in real-life multi-player analogues of *Prisoners' Dilemma*. Consider the decision to vote. In a large two-candidate election, all of the following are plausible. (a) The inconvenience of voting means that not voting is better than voting, even for a partisan agent, on either hypothesis about the outcome.[44] (b) If the expected turnout is in the millions then any agent should be practically certain that his vote will not *make* any difference to the outcome.[45] For instance, in the electoral college system governing US presidential elections, any urban voter in a non-marginal state like New York or Texas should be more confident of being run over on the way to voting. (c) Many agents take their own voting decision to be diagnostic of the outcome of large elections.[46]

---

[44] I am here assuming away any symbolic or expressive value that voting might hold (Riker and Ordeshook 1968; for criticism see Green and Shapiro 1994: 51–2). Although it is probably true that these things do matter to some voters, it is likely, especially in the light of evidence from Quattrone and Tversky (1986: see n. 46 in this chapter), that there are many for whom it is not decisive.

[45] Downs 1957: 246.

[46] E.g. Koestner et al. (1995) found that in the 1992 Canadian referendum on constitutional reform, students who planned to vote yes (to the proposed amendments) estimated on average that 56 per cent of the electorate would vote yes, whereas students who planned to vote no estimated that 51 per cent of the electorate would vote yes.

Now many people do in fact vote in spite of (a) and (b). And perhaps what gets them out of bed on polling day has something to do with (c).[47] If so, we have a multi-player analogue of a simulatory *Prisoners' Dilemma* in which a majority chooses to co-operate.[48] Chapter 7 addresses the question

[47] Quattrone and Tversky 1986; Grafstein 1991. Quattrone and Tversky asked 315 university students to imagine deciding whether to vote in an election in the fictitious state Delta. There are two parties in Delta, A and B, with about 4 million firm supporters each, including the subject, who supports Party A. In addition the electorate contains 4 million non-aligned voters. Quattrone and Tversky then presented each subject with one of two theories about the result of the election. According to the *Party Supporters Theory*, the margin of victory would be a few hundred thousand votes in one or the other direction, and non-aligned voters would split their vote evenly between the parties. So the turn-out amongst the 8 million committed supporters would be the decisive factor, on this theory. But according to the *Non-aligned Voters Theory*, party supporters would turn out in roughly equal numbers. So on this theory it is the non-aligned voters whose preferences would decide the election. The difference between these theories that matters is that a subject who (a) took his own vote as a sign of higher participation amongst members of his own group, and (b) followed the evidentialist prescription in this case, would take the Party Supporters Theory, but not the Non-aligned Voters Theory, as a reason to vote for Party A rather than to abstain. On the other hand a subject who regarded his own vote as diagnostically irrelevant to turnout amongst A-supporters, or who cared only about the *causal* impact of his vote on the results, which on either theory is negligible, would be equally likely to vote in the light of either demographic theory. Subjects were asked six questions. The first four asked each subject to estimate the probabilities of relative turnout amongst firm party supporters, and of overall victory for Party A, conditional on his own voting and on his own abstention. The fifth question asked each subject to assess the likelihood of his voting, given that voting in Delta is costly and in the light of whichever demographic theory he is being asked to entertain. The sixth question asked for a simple yes or no on the same matter. If a significant number of subjects subscribed to both of points (a) and (b) as stated above, then you would expect different distributions of answers amongst subjects who were given the Party Supporters Theory and amongst those who received the Non-aligned Voters Theory. In particular you would expect that amongst the first type of subject, the probability of more votes for A conditional on his voting (and also of victory for A conditional on his voting) would exceed the probability of more votes for A conditional on his *not* voting (resp. victory for A conditional on the same) by more than for the second type of subject. And you would expect to get answers (equivalent to) *high* and *yes* to the fifth and sixth questions significantly more frequently from amongst the first type of subject than from amongst the second type of subject. The experimenters did observe just these results. On the whole it seems that a partisan agent *is* prepared to go to the trouble of voting in elections where turnout for his party makes a difference to the result, even though he knows that his own vote could not possibly make any difference (all subjects knowing that the margin of victory would be at least 200,000). As Quattrone and Tversky write (1986: 54–5): 'From the perspective of the individual citizen, voting is both causal and diagnostic with respect to a desired electoral outcome. Causally, a single vote may create or break a tie, and the citizen may communicate with like-minded peers, persuading them also to vote. Diagnostically, one's decision to vote or to abstain is an indicator that others who think and act like oneself are likely to make the same decision. The Party Supporters and Non-aligned Voters theories were equivalent in the causal significance of voting. But subjects perceived the Party Supporters Theory as having more diagnostic significance than the Non-aligned Voters Theory. As a consequence, they indicated a greater willingness to vote given the validity of the former theory than given the validity of the latter. These results obtained despite the margin of victory's being kept at from 200,000 to 400,000 votes for both theories.'

[48] For instance, Huntingdon in Cambridgeshire, UK, has been a completely safe Conservative parliamentary seat ever since its creation in 1983. But in every general election since then, turnout in this seat has exceeded 60 per cent. This is in line with Shafir and Tversky's (1992: 462) finding that 65 per cent of subjects in their version of Newcomb's problem (see section 4.5) chose to one-box.

whether anything can be said for these intuitions. For now the point is only that they exist, and so the fact that *Prisoners' Dilemma* is a Newcomb problem is hardly decisive grounds for rejecting EDT in favour of CDT.

The overall conclusion of this chapter is that nearly all of the cases that causalists have taken to be realistic situations where EDT and CDT disagree are nothing of the sort. The remedial cases, and also the medical, economic and psychological versions of the Newcomb problem, are genuine cases of disagreement only at great cost to their realism, a cost that is so great as to make them irrelevant if realism is a concern. And if realism is not a concern, then we may as well focus exclusively on the standard Newcomb case, which is at least frankly science-fictional.

On the other hand and contrary to what some evidentialists seem to have thought,[49] the Tickle Defence constitutes no general strategy for ruling out a priori all possibility of a realistic Newcomb problem. On plausible assumptions, *Prisoners' Dilemma* constitutes a case that is realistic and involves genuine disagreement between EDT and CDT. But unlike the remedial and medical Newcomb cases, it is *not* a decision situation in which CDT straightforwardly and obviously gives the right advice. So it cannot straightforwardly establish causalism. In fact the disagreement over this case may turn out to threaten causalism.

The next two chapters develop this theme in connection with some new cases of disagreement. These are unrealistic in the sense of being unlikely to arise in real life, but realistic in the senses of being practically feasible and of not involving far-fetched beliefs. If, as I'll argue, EDT gives *better* advice than CDT in these cases, this relative degree of realism means that they should weigh more heavily against causalism than any conclusion about the standard Newcomb problem might weigh in its favour.

Similarly, when Gardner (1974: 102) published Newcomb's problem in *Scientific American*, approximately 70 per cent of the very many readers who wrote in expressing an opinion chose to one-box (*Scientific American*, March 1974: 102). Anand (1990: 64) achieved similar results amongst subjects who thought, or at least were told, that the predictor was highly reliable.

[49] Horwich 1987: 179–85; Price 1991: section 4.

CHAPTER 5

Deterministic cases

This chapter considers two cases that appear to count against CDT. Although neither case trades solely on such widespread and common-sensical beliefs as those with which their proponents sought to motivate the remedial, medical and other cases that Chapter 4 discussed, neither requires such an outlandish suspension of belief as we find in the standard Newcomb problem. In both of these cases, I ask the reader to accept a reasonable but far from mandatory theoretical assumption concerning the large-scale workings of the universe. CDT turns out inconsistent with this assumption.

5.1 *Betting on the Past*

In this section, and then in section 5.2, Alice is going to face two separate decision problems. Here is the first.

> *Betting on the Past*: In my pocket (says Bob) I have a slip of paper on which is written a proposition *P*. You must choose between two bets. Bet 1 is a bet on *P* at 10:1 for a stake of one dollar. Bet 2 is a bet on *P* at 1:10 for a stake of ten dollars. So your pay-offs are as in Table 5.1. Before you choose whether to take Bet 1 or Bet 2 I should tell you what *P* is. It is the proposition that the past state of the world was such as to cause you now to take Bet 2.

Let soft determinism be the doctrine that: (i) Alice is free now to take either bet, and (ii) at every past time there already existed a determining cause for her now taking whichever bet she actually *does* take. Suppose that Alice is highly confident of soft determinism and certain of everything that Bob tells her. Should she take Bet 1 or Bet 2?

Causal Decision Theory (CDT) recommends that Alice take Bet 1. But she *should* take Bet 2. So Causal Decision Theory is false. Or so I'll argue. But let me first describe things more carefully.

Table 5.1 *Betting on the Past:*
approximate version

	P	$\neg P$
O_1: take Bet 1	10	-1
O_2: take Bet 2	1	-10

5.1.1   *The case in more detail*

My description of Alice's case is completely vague about (a) the betting mechanism, vague and slightly inaccurate about (b) soft determinism, and slightly inaccurate about (c) the content of P. Let me now state (a) and (b) more precisely and (b) and (c) more accurately.

(a) The betting mechanism. Alice has a choice between now raising her hand and now doing something else. Suppose that she will in fact either *raise* her hand or *lower* her hand and is certain of this. Then it seems to Alice that she has a choice between now raising her hand and now lowering her hand.

Suppose *raising* her hand signals to Bob that she accepts Bet 1, and *lowering* her hand signals that she accepts Bet 2. So her choice between raising her hand and lowering her hand *is* her choice between Bet 1 and Bet 2. Again she is certain of all of this.

(b) Soft determinism. I just called soft determinism the view that Alice is free to take either bet, and that at every past time there already existed a determining cause for her now taking whichever of Bet 1 and Bet 2 she actually *does* take.

Talk of 'determining causes' helps convey the general idea. But we can explain soft determinism more precisely and accurately without it. Soft determinism says this about Alice's present betting behaviour:

(5.1)   **Soft determinism:** (i) Alice does freely whatever she will actually do. (ii) There is a true historical proposition H about the intrinsic state of the world long ago, and there is a true proposition L specifying the laws of nature that govern @, such that H and L jointly determine what Alice will do.[1]

---

[1]   Cf. Lewis 1981a: 291–2. Lewis's own example concerns what the agent *did*, not what the agent *will* do. But I cannot see that I am straying from Lewis's reading of soft determinism. If I am, then Lewis must be holding that determinism says that past acts of mine *were* determined by events in their pasts but that future and present acts *are* not, or *will* not be, determined by events in *their* past. I am

Proposition (5.1)(i) is the soft part. Proposition (5.1)(ii) is the determinism part. '@' rigidly designates the actual world.

I said that Alice is highly confident of soft determinism because she is certain of the soft part and very nearly certain of the determinism part. More specifically, she places nearly all of her credence about the laws of nature in a specific system L^* that is deterministic in the sense of (5.1)(ii) and which explains all of the anomalies that faced earlier theories. It reconciles quantum theory and general relativity. Stringent experiments have repeatedly confirmed it. It has no serious competitors, and so on. But she needn't be *completely* certain of L^*. Suppose instead that her $Cr(L^*) = 1 - \Delta$ for some minute positive Δ. And to repeat, she knows that L^* is deterministic in the sense that (5.1)(ii) spells out.

But that spelling out itself bears further explanation, as does (5.1)(i). More specifically, I should explain four expressions in (5.1): 'Alice does *freely* whatever she will actually do', 'about the *intrinsic state* of the world long ago', 'the *laws of nature* that govern @' and 'H and L *jointly determine* what Alice will do'.

I can't define 'Alice does freely whatever she will actually do' in any way that satisfies all our intuitions about 'free'. But I won't need to. My argument only cares about soft determinism because of its consequences for decision theory. So I only need to give a *necessary* condition for 'Alice does freely whatever she will actually do' that connects (5.1) with the latter. The relevant necessary condition is that Alice is free in the sense that CDT *should tell her* what to do. So if CDT is true then it *does* tell her what to do.

This is plausible. For if you are free to do A, and free to do B instead, then there is some point in *deliberating* whether to do A or B. And the true normative decision theory, whatever it is, will *tell* you which of A and B to do if there *is* some point in deliberating which of A and B to do. So if CDT is true then it will tell Alice what to do. So I propose:

(5.2)   **Freedom**: Alice does freely whatever she will actually do only if CDT should tell her what to do.

Lewis once defined an intrinsic property of a *region* as one such that, whenever two possible regions are perfect duplicates, the property belongs to both or neither.[2] And he defined a proposition as being *about* a subject

---

not even sure that that position is diachronically consistent. Consistent or not, it leaves the future undetermined, so doesn't deserve to be called determinism at all.

[2] Lewis 1986: 263. Lewis notes (ibid. n. 14) that here 'we are dealing with a substantial circle of interdefinables, and so have a choice of alternative primitives'.

matter if and only if any two possible worlds that agree on the subject matter agree on the proposition.[3] Combining these, let us try:

(5.3)   **About the intrinsic state of the world at time *t*:** A proposition *H* is about the intrinsic state of the world at time *t* if and only if, whenever two possible worlds w_1 and w_2 are perfect duplicates at *t*, *H* is either true at *both* of w_1 and w_2 or true at *neither*.

One might object to the consequence of (5.3) that all necessary truths and falsehoods are about the intrinsic state of the world at all times. That may be counterintuitive. But it is harmless for *my* purposes, which involve only *contingent* historical propositions.

Relatively little turns upon how exactly we explain 'the laws of nature that govern @'. Lewis's account serves perfectly well:

(5.4)   **Laws of nature:** The laws of nature that govern @ are those regularities that would come out as axioms in a system that was optimal among actually true systems in its combination of simplicity and strength.[4]

One consequence of this definition that *will* be important is that it allows what (5.1) requires, namely that the laws are *true*: in fact it entails that. So Alice knows that any law of our world tells the truth about it.[5]

As for 'determine', we might try this: *H* and *L* determine what Alice will do if and only if: *either* it is deducible from *H* and *L* that Alice will *raise* her hand, *or* it is deducible from them that she will *lower* her hand.

But there are two objections to this. First, determination as in (5.1)(ii) clearly relates *H* and *L* to possible *events*, not to sentences. Second, no sensible soft determinist would commit herself to a truth describing movements of Alice's hand that is deducible from *H* and *L*. After all, *L* probably makes no mention of macroscopic bodily movements at all. It is more likely that it describes e.g. distributions of gravitational and electromagnetic potential at times or the evolution of these distributions over time. So probably *no* sentence describing the motion of Alice's hand is *deducible* from *H* and *L*.

---

[3] Lewis 1980: 93.

[4] Lewis 1979a: 55. Furthermore: (a) '@' and 'actually' are my interpolations. I am assuming that the result is still faithful to Lewis. (b) Proposition (5.4) is actually Lewis's definition of *fundamental* laws. The other laws, the *derived* laws, are the other theorems of that optimal system of which the fundamental laws are axioms. Nothing hangs on this. Soft determinism about fundamental laws is equivalent to soft determinism about (fundamental laws + derived laws). (c) See Lewis 1980: 123 ff. for more on 'simplicity and strength'.

[5] These Lewisian formulations of intrinsicness and lawhood suffice for my purposes but they are not necessary for them. The argument should go through on any formulations that entail (a) that the laws are exceptionless if determinism is true and (b) that intrinsic states of the distant past are causally independent of Alice's present actions.

I prefer to explain determination in terms of *supervenience* of *events*:

(5.5) **Determination:** Propositions $P_1, P_2, \ldots P_n$ jointly *determine* a possible event E if and only if E occurs at every possible world at which $P_1, P_2, \ldots P_n$ are true.[6]

According to (5.1)–(5.5) then, Alice's soft determinism is soft because she thinks CDT should tell her *whether* to raise her hand.[7] And it is deterministic because she thinks that for some historical proposition H, true at all worlds that duplicate the state of @ at a distant time t, and some proposition L that states the laws of @, any world where both are true is one where Alice does whatever she *actually* does. So if she raises her hand at @ then she raises her hand at *every* world where H and L are both true. If she lowers her hand at @ then she lowers her hand at *every* world where H and L are both true.

(c) The content of P. In the original story P says that something happened *at some time* in the past to *cause* Alice now to *take Bet 2*. In fact P is more specific about the time that it describes, it doesn't mention causation at all but rather the specific system L^*, and it mentions Alice's bet in terms of a bodily motion.

Let T be some particular time in the distant past, say Christmas Day, 10 million BC. Then what P says is this:

(5.6) **The proposition P:** There is a true historical proposition H describing the intrinsic state of the world at T, such that H and L^* jointly determine that Alice now lowers her hand.

So P is quite specific about the time that it describes: Christmas Day, 10 million BC. P doesn't mention causation but only laws of nature, as explained at (5.4). And P doesn't actually describe Alice's bet on or against P itself in such terms. It only states that back then at T, there existed L^*-antecedents for Alice's now lowering her hand. So it is possible to explain the content of P without mentioning P itself. P also involves the concepts of

---

[6] Does this formulation contradict my assertion, at section 1.2, that we needn't think of possible worlds as concrete entities but could instead treat them simply as elements of propositions? No. To say that 'an event E occurs at a world' need not itself be interpreted as saying that something concrete is going on at some non-actual sector of reality but rather that a proposition *describing E* in suitable terms is true there. There is still no requirement that the proposition be deducible from $P_1, P_2, \ldots P_n$.

[7] Note that 'CDT should tell her whether to raise her hand' here means 'If CDT is true then it *does* tell her whether to raise her hand'. Clearly this doesn't suffice for softness in the usual sense of 'soft determinism', since *it* doesn't entail that Alice is free, i.e. that she is in fact in a position to raise her hand and also to lower her hand. But the argument does not trade on any necessary *and* sufficient condition of Alice's being free, but only on the necessary condition that CDT if true must be giving her advice. If she is certain of this then she is as soft as necessary.

Table 5.2 *Betting on the Past:*
precise version

	P	$\neg P$
O_1: raise hand	10	-1
O_2: lower hand	1	-10

intrinsicality and joint determination. You should read these as explained at (5.3) and (5.5) respectively.

Finally, note especially that the truth-value of P at a world supervenes on its intrinsic state at time T. This is because P is true at any world w that contains at T nomological antecedents of Alice's now lowering her hand *according to L^**, even if she *doesn't* lower her hand *at w* because L^* doesn't describe the laws there. Suppose, for instance, that L^* is such that any L^*-world at which electron e is in state S on Christmas Day, 10 million BC, is a world at which Alice now lowers her hand. Then P is true at *every* world w at which electron e is in state S on Christmas Day, 10 million BC, even if Alice doesn't exist at w. (From now on **P-world** means 'world at which P is true' etc.)

In light of all this, let me now state the argument in more detail. What Bob actually says to Alice is this:

> Raising your hand commits you to betting one dollar at 10:1 on the truth of the proposition P as stated at (5.6). Lowering your hand commits you to betting ten dollars at 1:10 on the truth of P. Your options, my abbreviations for them, the relevant events and the pay-offs are as in Table 5.2. How do you want to bet?

The argument that this case refutes CDT is as follows:

(5.7) **Relevance premise:** If CDT is true then it gives Alice correct advice.

(5.8) **Descriptive premise:** CDT advises Alice to realize O_1.

(5.9) **Normative premise:** Alice should realize O_2.

(5.10) **Conclusion:** CDT is false

The argument is valid, and the relevance premise (5.7) follows straightforwardly from (5.1)(i) and (5.2). It therefore remains to defend the descriptive premise (5.8) and the normative premise (5.9).

5.1.2 The descriptive premise

It follows from (5.6) that P is about the past in the sense of (5.3). It is therefore true that the partition $\{P, \neg P\}$ is causally independent of anything that Alice now does. It then follows straightforwardly from (2.34) that:

(5.11) $\quad U(O_1) = 10\,Cr(P) - Cr(\neg P)$
(5.12) $\quad U(O_2) = Cr(P) - 10\,Cr(\neg P)$

Since either $Cr(P) > 0$ or $Cr(\neg P) > 0$, it follows from (5.11) and (5.12) that $U(O_1) > U(O_2)$: CDT recommends that Alice raise her hand in order to bet that she was determined to *lower* her hand.

This very quick argument skates over a difficulty arising from the interaction of determinism and chance. After all, you might think that if L^* is true, then the past has already determined Alice to do one thing or the other. So either $Ch(O_1) = 0$ or $Ch(O_2) = 0$. So either $Ch(P \mid O_1)$ or $Ch(P \mid O_2)$ is undefined. In either case CDT gives no advice since the cells of the relevant K-partition are not all defined.

If that *were* true it would be bad news for CDT in light of the relevance premise (5.7), which follows from Alice's soft determinism. So this way of denying the descriptive premise does nothing to help CDT. But in any case it is possible to make sense of chances conditional on an option that is already causally determined not to be realized, for instance by defining the chance of P conditional on (say) O_1 in that case as the chance that P *would* have at the time of O_1 if O_1 *were* to be realized. We assess this counterfactual by imagining a world exactly like ours up to now, at which point O_1 comes true by some miracle. Clearly, since P is as true of the past of that world as it is of the past of ours, the chance that P is true is 1 if P is actually true and 0 if P is actually false: (2.34) then applies straightforwardly.

Another way around the difficulty would be to suppose that Alice places *all* of her credence not in L^* itself, but in some theory L^{**}, which says that at all times there is a very small chance Δ that L^* itself will be violated at some place, perhaps through some extraordinary form of quantum tunnelling: the sort of thing that would have, say, a 50 per cent chance of happening once in 10^{15} years. (But let P continue to speak of L^*.) Since L^{**} does not force either $Ch(O_1)$ or $Ch(O_2)$ to zero, the relevant conditional chances are meaningful and (2.34) again applies straightforwardly.

But these manoeuvres obscure the fact that my (perhaps oversimplified) initial argument was meant to bring to light, namely that any theory that respects the basic point behind CDT *must* recommend O_1 over O_2. CDT was supposed to capture the intuitive thought that you should realize the option that you expect causally to *make* you best off, regardless of what doing so might reveal about any of its *non*-effects that might also matter to you but which are beyond your present influence. And although Alice's present action does reveal something about the past (viz which of P and

$\neg P$ is true), it has no effect on the past. Since on either relevant hypothesis about the past (P, $\neg P$), Alice is better off with O_1 than with O_2, CDT should recommend O_1 over O_2 for essentially the same reason that it recommends two-boxing in the standard version of the Newcomb problem.[8]

5.1.3 The normative premise

It is difficult to *argue* that O_2 is rationally superior to O_1 because it seems intuitively so obvious that it is. And nearly everyone on whom I have tried this example agrees that it is. But let me briefly underscore that intuition.

Given her confidence $1 - \Delta$ in L^*, Alice *is arbitrarily confident* that anyone who lowers her hand in her situation was determined by L^* and H to lower her hand, and that anyone who *raised* her hand in this situation was *not* determined by L^* and H to lower her hand.[9] So she is practically certain in advance that if she lowers her hand now then P is true, and that if she raises her hand then $\neg P$ is true. From her perspective, anyone who lowers her hand in her situation is almost *certain* to make a one-dollar *profit*, and anyone who raises her hand in that situation is almost *certain* to make a one-dollar *loss*. What better reason could Alice have to lower her hand?

That is what EDT recommends. Given Alice's soft determinism her credences satisfy $Cr(P|O_1) \approx 0$ and $Cr(P|O_2) \approx 1$. So she has $V(O_1) \approx -1$ and $V(O_2) \approx 1$. So $O_2 \succ O_1$, and EDT recommends O_2 over O_1. Intuitively, this is because lowering her hand reveals she has won her bet

---

[8] A decision theory that does not respect this basic intuition, despite calling itself 'causal', is that of J. Cantwell (2013: 675–9). According to it, the truth of a deterministic L^* implies that there are only two relevant possibilities: either $O_1 \wedge \neg P$ is true, or $O_2 \wedge P$ is true. Accordingly, if Alice is very confident of L^* then she should choose O_2 over O_1. But in the indeterministic case where Alice is certain of L^{**}, Cantwell's theory recommends O_1, which is what standard CDT recommends in both cases. The reason for not calling this a 'causal' theory is that according to it, the rational choice between O_1 and O_2 does not depend only on their causal efficacy in bringing about what the agent wants. In a deterministic L^*-world and in an indeterministic L^{**}-world, O_1 and O_2 have the same effects, but Cantwell's theory makes different recommendations at the two worlds. A further peculiarity of Cantwell's theory should also make causalists uncomfortable: it implies that the right choice in any Newcomb problem depends on your confidence that determinism is true. If you are confident that determinism is false then you should take both boxes. But as your confidence in determinism rises the case for one-boxing becomes increasingly strong. At some point well before your confidence in determinism reaches certainty, the theory abruptly switches to recommend two-boxing. But the causalist case for two-boxing in Newcomb's problem is independent of whether determinism is true.

[9] This remains true when applied to the version of the case involving the indeterministic L^{**} mentioned briefly at section 5.1.2. That is: given her complete confidence in L^{**}, Alice is arbitrarily confident that anyone who lowers her hand in her situation was determined by L^* and H to lower her hand, and that anyone who *raised* her hand in this situation was *not* determined by L^* and H to lower her hand. So the argument also works against Cantwell's version of CDT (see n. 8 above), which recommends O_1 in this indeterministic version of the example.

and raising her hand reveals that she has lost it. More generally, the example exploits the fact that a soft determinist must take any action of hers always to reveal something to her about the distant past. It reveals that the state of the world at that past time was such as to cause her now to realize just *this* option.[10]

But for the causalist this is all nonsense. Either the state of the world ten million years ago was such as to L^*-determine Alice now to lower her hand, or else it was not. But in either case it is in the past. There is nothing that she can do about it now. So if P is true then she has this choice: raising her hand and getting ten dollars versus lowering her hand and getting one dollar. And if P is false then she has this choice: raising her hand and losing one dollar versus lowering her hand and losing ten dollars. In either case raising her hand makes her nine dollars better off, so she ought to do that.

Just to be clear about what this argument is: the premise 'There is nothing that she can do about it now' had better not *just* mean that either it is already true that P or it is already true that $\neg P$. *That* tautology is true whether P is about the past or the future, because just as what's done is done, so too what will be will be. Thus in *Sink or Swim* (Table 1.4), either it is already true that Alice will drown or it is already true that Alice will not drown. But all non-fatalists agree that the fact, that in either case she is better off not learning to swim, does nothing to show that she should in fact *not* now learn to swim. 'There is nothing that she can do about it now' must mean: nothing that she does now can *causally* affect whether P is true or not.

This again raises the general question that came up at section 4.6.3: what is it *about* causal independence that *makes* for a difference between legitimate and illegitimate applications of the dominance principle? In particular, why is it causal independence, and not *nomological* independence, that matters? The contrast between causal and nomological independence is what makes the difference between causalist and evidentialist attitudes towards *Betting on the Past*. P's truth is causally independent of what Alice now does, but it is not *nomologically* independent: in fact she is very confident that the laws guarantee that she will raise her hand if and only if P is false. The causalist argument for O_1 therefore turns on the issue between these two principles:

(5.13) **Causal dominance**: The dominance principle applies whenever the relevant events are causally independent of the agent's options.

---

[10] No tickle-style screening-off is available in this case. Even if Alice has perfect knowledge of her present beliefs and desires, still she doesn't whilst deliberating know *which* option they will in fact cause. Hence it is only after she has decided that she can decisively settle which of P and $\neg P$ is true. *Ex ante* therefore, nothing short of her actual choice screens off the latter from P.

(5.14) **Nomological dominance:** the dominance principle applies only when the relevant events are nomologically independent of the agent's options.

If (5.13) is true then we should accept the causalist argument that Alice should raise her hand. If (5.14) is true then we should reject it. How do we settle this?

I have not been able to find any grounds for asserting the causal dominance principle (5.13) or for its specific application to the present case. The standard practice amongst causalists is simply to *assert* it.[11] On the other hand, we have the best possible grounds for accepting nomological dominance, at least as applied to Alice's current difficulty, namely *empirical* evidence that it *works*. Specifically: we have excellent empirical grounds for L^*, which implies that anyone who applies the dominance principle in *Betting on the Past* will *lose* one dollar, and anyone who rejects it will *make* one dollar. So we have strong empirical grounds for thinking that out of (5.13) and (5.14), only the nomological principle is compatible with the better strategy.

More generally, suppose that we have:

(a) strong empirical grounds for a nomological connection between options $\{Q_1, Q_2\}$, and events $\{S_1, S_2\}$ forming a partition of S that is rich given an element of $\{Q_1, Q_2\}$, such that $Q_1 \equiv S_1$ and $Q_2 \equiv S_2$ are nomologically guaranteed;
(b) preferences $\succ$ over outcomes such that $Q_1 \wedge S_1 \succ Q_2 \wedge S_2$.

Then the laws of nature themselves give strong empirical grounds to expect that Q_1 works better than Q_2, in the sense that agents that realize Q_1 will do better than agents that realize Q_2. We know what will work because we have empirical evidence that it will work, and no amount of philosophizing is going to change that. If a causalist wishes to reject the force of this argument for the evidentialist strategy in *Betting on the Past*, or any other case that opposes nomological dependence and causal independence, he must say exactly how this argument falls short of being an empirical vindication of that strategy.

After all, such empirical vindications seem perfectly adequate for all practical purposes. If you know, by observation, testimony or theory, that

---

[11] E.g. Gibbard and Harper's (1978: 361) defence of their version of (5.13) is that it seems to them mistaken to ignore the good effects of an option because it is not on balance a good sign. This just repeats a preference for CDT over EDT. Similarly Nozick (1969: 219–26); Skyrms (1980: 128–30); Lewis (1981b: 309); Joyce (1999: 150–1); Weirich (1998: 116–7; 2001: 126); and Sloman (2005: 90) simply repeat the causal dominance principle with varying degrees of emphasis and circumlocution but in all cases without supporting argument. Pollock (2010: 64) points out that the appeal to this principle begs the question in the context of the standard Newcomb problem. What he says there applies equally to *Betting on the Past*.

anyone who touches the fire gets burnt, then that is good enough reason not to touch the fire. You needn't know or care whether touching the fire *causes* you to get burnt. You needn't even – and if you are a child or a primitive you won't – have any *concept* of causation. It is still the case that repeatedly getting burnt by the fire is already reason not to put your hand in the fire, because it is reason to think that fire burns (that is, to expect that anyone who touches the fire gets burnt). Similarly then, Alice knows by observation, testimony or theory that following CDT makes her one dollar worse off in *Betting on the Past*. Why isn't that at least as good a reason not to follow CDT as her knowledge that fire burns was not to touch the fire?

This argument for nomological over causal dominance is a special case of a more general argument for evidentialism. That is the 'Why Ain'cha Rich?' argument that section 7.3 discusses. Unless the causalist can find something wrong with it (and I'll argue there that he cannot), it would be misleading to describe the disagreement between EDT and CDT over *Betting on the Past* as a stand-off.[12] Causalist stubbornness is not the same thing as a genuine stand-off. The fact is that in *Betting on the Past*, the evidentialist can point not only to intuitive but also to empirical grounds for O_2, grounds that are at least as strong as Alice's evidence for L^*. The causalist can point to nothing.

5.2   *Betting on the Laws*

The next day Alice faces a second problem.

> *Betting on the Laws*: On this occasion and for some irrelevant reason, Alice must commit herself in print to the truth or falsity of her favoured system of laws L^*. She has two options: O_1: to *affirm* L^*; that is, to assert that our world in fact *conforms* to the (jointly) deterministic generalizations that it conjoins; and O_2: to *deny* it; that is, to assert that at some time and place our world *violates* at least one of these generalizations.

Since Alice is a philosopher, she cares what she says because, and only because, she wants to tell the truth. Her V-scores for the possible outcomes in this situation are therefore as in Table 5.3.

The argument at this point proceeds on the same broad lines as at section 5.1:

(5.15)   **Relevance premise:** If CDT is true then it gives Alice correct advice in *Betting on the Laws*.

---

[12] This appears to be the position of Lewis, who describes the situation as deadlocked (1981b: 306) without mentioning any flaw in the empirical argument as applied to Newcomb's problem (1981c).

Table 5.3 *Betting on the Laws*

	L^*	$\neg L^*$
O_1: affirm L^*	I	0
O_2: deny L^*	0	I

(5.16) **Descriptive premise:** CDT endorses O_2 in *Betting on the Laws*.

(5.17) **Normative premise:** O_1 is uniquely rational in this problem.

(5.18) **Conclusion:** CDT is false.

Proposition (5.15) follows from (5.1)(i) and (5.2), which we are not suspending. And the argument is valid. So it remains to defend premises (5.16) and (5.17).

5.2.1 The descriptive premise

There is a simple intuitive case that CDT endorses O_2. Say that a possible world is within Alice's causal reach just in case it differs from @ only over particular matters of fact that are causally dependent on what Alice now does. Then the basic point of Causal Decision Theory is that the only possible worlds that matter for assessing O_1 and O_2 are those that are within Alice's causal reach.

Any past particular matter of fact is causally independent of what Alice now does. So all the possible worlds that are within Alice's present causal reach match @, and so also match one another, over all particular past matters of fact. From the worlds within her causal reach, select any O_1-world w_1 and any O_2-world w_2. Since w_1 and w_2 have identical pasts but differing presents, one of the following is true:

(5.19) w_1 is an L^*-world and w_2 is a $\neg L^*$-world

(5.20) w_1 is a $\neg L^*$-world and w_2 is an L^*-world

(5.21) w_1 and w_2 are both $\neg L^*$-worlds

Whichever of these hypotheses is true, Alice is at least as well off in w_2 as she is in w_1. If (5.19) is true, she wins her 'bet' at w_1 and at w_2. If (5.20) is true then she loses at both worlds. And if (5.21) is true then she loses at w_1 and wins at w_2. Since w_1 and w_2 were arbitrary, it follows that at any O_2-world within her causal reach she does at least as well as at any O_1-world within her causal reach. So anything worth calling *Causal* Decision Theory had better endorse O_2.

Applying this to our official version of CDT does not complicate things much. Let Q be some true proposition that is (in the sense of (5.3)) wholly about the intrinsic state of the world in the distant past; let Q be so detailed as to L^*-determine its future. So any Q-worlds that disagree post-Q are not both L^*-worlds. (If we like, we can say that Q is the set of possible worlds that match @ over all particular matters of fact on Christmas Day, 10 million BC.) Then whether Q is true is causally independent of what Alice now does; since Q concerns the past, we therefore have:

(5.22) $Ch\,(Q\,|\,O_1) = Ch\,(Q\,|\,O_2) = 1$

Now two simple theorems of the probability calculus are that:

(5.23) $Ch\,(Q\,|\,O_1) = Ch\,(Q\,|\,L^* \wedge O_1)\,Ch\,(L^*\,|O_1) + Ch\,(Q|\neg L^* \wedge O_1)$
          $Ch\,(\neg L^*\,|O_1)$

(5.24) $Ch\,(Q\,|\,O_2) = Ch\,(Q\,|\,L^* \wedge O_2)\,Ch\,(L^*\,|O_2) + Ch\,(Q|\neg L^* \wedge O_2)$
          $Ch\,(\neg L^*\,|O_2)$

Since Q either L^*-determines O_1 or L^*-determines O_2, one of the following is true:

(5.25) $Ch\,(Q\,|\,L^* \wedge O_1) = 0$

(5.26) $Ch\,(Q\,|\,L^* \wedge O_2) = 0$

And given the disjunction of (5.25) and (5.26), it follows from (5.22)–(5.24) that one of the following is true:

(5.27) $Ch\,(\neg L^*\,|O_1) = 1$

(5.28) $Ch\,(\neg L^*\,|O_2) = 1$

But in either of these cases $U\,(O_2) \geq U\,(O_1)$: in case (5.27) because O_1 makes Alice sure to lose, and in case (5.28) because O_2 makes her sure to win. So $U\,(O_2) \geq U\,(O_1)$: CDT recommends betting that L^* is *false*.[13]

One objection to this argument is that the chance function gives the laws a chance of 1; so if L^* is true then either $Ch\,(O_1) = 0$ or $Ch\,(O_2) = 0$, and so one of the marginal chance functions $Ch\,(x \mid O_1)$ and $Ch\,(x \mid O_2)$ is (very probably, Alice should think) undefined. In that case CDT makes no recommendation about the case, and that in itself spells trouble for this version of CDT in light of the relevance premise (5.15).

As briefly discussed at section 5.1.2, a version of CDT that clearly does make a recommendation to Alice is a counterfactual version according

---

[13] This statement of the argument is something of a simplification. The relevant hypotheses (elements of the K-partition) are the elements of $K_1 = \{Ch\,(\neg L^* \mid O_1) = 0) \wedge Ch\,(L^* \mid O_2) = x \mid 0 \leq x \leq 1\}$ and of $K_2 = \{Ch\,(\neg L^* \mid O_1) = x \wedge Ch\,(\neg L^* \mid O_2) = 0 \mid 0 \leq x \leq 1\}$. So $U\,(O_1) = \int_{0 \leq x \leq 1} x\,dF$, where $F\,(x) = Cr\,(Ch\,(L^* \mid O_1) \leq x)$; and $U\,(O_2) = 1 + \int_{0 \leq x \leq 1} x\,dG$, where $G\,(x) = Cr\,(Ch\,(\neg L^* \mid O_2) \leq x)$. Writing $f = F'$ and $g = G'$, we have $U\,(O_1) = \int_0^1 x f(x)\,dx$ and $U\,(O_2) = 1 + \int_0^1 x\,g\,(x)\,dx$. Since $F\,(x)$ and $G\,(x)$ are non-decreasing for $0 \leq x \leq 1$, $f(x) \geq 0$ and $g\,(x) \geq 0$ for these values of x. Since $g \geq 0$, $\int_0^1 x\,g\,(x)\,dx \geq 0$. So $U\,(O_1) \leq \int_0^1 f(x)\,dx = 1 \leq 1 + \int_0^1 x\,g\,(x)\,dx = U\,(O_2)$.

to which Alice should evaluate the options O_1 and O_2 in the light of her confidence that she would be right if she were to affirm L^*, and her confidence that she would be right if she were to deny it. This formulation of causalism certainly does apply to *Betting on the Laws*; but it is easy to see that it endorses O_2.[14]

Another way to make something like CDT apply to this case would be to calculate the conditional chance functions only where these are defined. In the present case, the upshot of that would be that:

(5.29) $U(O_1) = \int_{0 \leq x \leq 1} x \, dF$

(5.30) $U(O_2) = \int_{0 \leq x \leq 1} x \, dG$

where $F(x) = Cr(Ch(L^* | O_1) \leq x \mid Ch(O_1) > 0)$ and $G(x) = Cr(Ch(\neg L^* | O_2) \leq x \mid Ch(O_2) > 0)$. Since Alice is very confident that L^* is true and so has a chance of 1, it follows that $U(O_1) \approx 1$ and $U(O_2) \approx 0$. So $U(O_1) > U(O_2)$, and now this *quasi-CDT* seems to recommend that Alice *affirm L^**.

But in taking this approach to deterministic cases, quasi-CDT has lost contact with the causalism that motivated CDT in the first place. Informally, what we are now being asked to compare is how well Alice does in O_1-worlds and O_2-worlds that largely differ *over their pasts*. The assessment of O_1 now depends on Alice's pay-off in an O_1-world w_1 that conforms to L^* and in which the past L^*-determines O_1; the assessment of O_2 depends on Alice's pay-offs in an O_2-world w_2 that conforms to L^* and in which the past L^*-determines O_2 instead. But w_1 and w_2 must disagree over the past. For instance, exactly one of them is a Q-world. In short: we are treating as relevant, for the assessment of Alice's present options, their pay-offs at worlds that are not in her present causal reach.

A further symptom of this is the fact that quasi-CDT would endorse one-boxing in deterministic versions of Newcomb's problem and the dominated strategy in deterministic versions of *Prisoners' Dilemma*, for instance in the Laplacean example (ii) at section 4.6.3. Whatever the actual plausibility of one-boxing in a deterministic Newcomb problem, it is clear enough that the only way to justify it would be to value it by something other than its causal effects.

I therefore conclude that this second amendment of CDT, which prefers O_1 over O_2, is not really a *Causal* Decision Theory at all. What these difficulties illustrate is not so much that there is any question whether causalism should endorse O_2, but only the awkwardness of applying a chance-based version of CDT to deterministic cases such as *Betting on the*

---

[14] This counterfactual version of CDT is proposed and defended in Gibbard and Harper (1978). For an argument that it endorses O_2 see Ahmed (2013: 294–6).

Laws. In such cases, the point stands out more clearly in connection with the counterfactual version of the theory, to which I shall have occasion to return at section 5.3.3.

5.2.2   The normative premise

There is a strong case that Alice should affirm L^* and should *not* deny it. We know that L^* is better confirmed than any of its competitors. The grounds for this confirmation have nothing to do with whether or not Alice bets on it, and nothing that she is now in a position to do will do anything to undercut those grounds, either for her or for anyone else. Similarly, there may be some crucial future experiment that finally settles once and for all whether L^* is true, but then nothing that she does now will have any effect on our confidence in the outcome of that experiment. In short, L^* is a truth that it is rational for Alice to bet on if anything is rational for her to bet on.

Of course Evidential Decision Theory respects all of these facts. Since $Cr_{ALICE}(L^* \mid O_1) = Cr_{ALICE}(L^* \mid O_2) > 0.5$, it follows straightforwardly from (2.18) that $V(O_1) > V(O_2)$: EDT agrees with common sense on this; and the empirical reasons that I cited at section 5.1.3 speak for it here too.

But Causal Decision Theory *cannot* agree with common sense. All of the facts that I just cited cut no ice against what is, for it, the decisive point that Alice's choice of bet is causally relevant to whether L^* is true, as argued in the previous section. Here the focus on causal dependence is counter-productive: when deciding whether to bet on or against a deterministic system of laws, you should ignore the fact that the truth of that system is causally dependent on what you ultimately do.

This difficulty therefore reveals a connection, at which the CDT literature occasionally and darkly hints, between that doctrine and incompatibilism about free will.[15] Certainly it is true that no incompatibilist will ever face the problems that face Alice, because no incompatibilist who thinks Alice is free will give much credence to a deterministic system like L^*. But that is not much comfort to the causalist who thinks, as I do, that there is nothing wild or incoherent about soft determinism – in fact it is far more plausible than what you are supposed to believe in the standard version of the Newcomb problem – and hence that CDT *should* be giving advice to people who believe it and want to act on its basis.

---

[15] Meek and Glymour 1994: 1007–8; Hitchcock 1996: 521–2.

5.3 Objections

Although the arguments in this chapter concern different cases and proceed from different premises, they invite related objections and raise common issues. This section considers three such objections. Section 5.4 states the common bearing of both arguments upon a matter of wider metaphysical significance.

5.3.1 The agency theory of causation

One might deny premise (5.8) of the first argument on the grounds that CDT only *seems* to recommend O_1 in *Betting on the Past* because it only *seems* that whether or not P is true is causally independent of what Alice does. On the correct theory of causation this is *not* the case.

I have in mind the theory that defines causality by its relevance to control and manipulation: causes are causal *handles*. That *agency* theory makes an agent take B to be causally dependent upon A whenever the evidential connection between A-type and B-type events survives the hypothesis that A is open to the agent's direct manipulation.[16]

To give an idea of its content: the theory implies that all genuine Newcomb problems involve temporally backwards causation. In these cases the evidential connection, between your taking the transparent box and the prediction that you will do this, survives the hypothesis that the former is under your direct control. So the *later* choice retrocausally influences the *earlier* prediction. So the agency theory makes CDT agree with EDT that you should one-box in the standard version of the Newcomb problem.

It also implies that Alice's action is causally relevant *to*, as well as being caused *by*, the state of @ in the distant past. So (2.34) does not apply, and (5.11) and (5.12) fail. So too therefore does my argument for (5.8).

But our everyday causal concept probably has the shape that it does because of *several* independent pressures. The need to mark out what is open to manipulation is certainly one of them. But the need to mark out certain types of continuous or quantity-preserving process is another, the need to make predictions is a third, and the need to give explanations is possibly a fourth. On this 'family resemblance' view of causation, an account of causation that focuses on just one of these pressures is inevitably going to distort our actual concept.[17]

There are other reasons for dissatisfaction with the agency theory. If it is going to save CDT from preferring O_1 in *Betting on the Past* or anything

---

[16] Menzies and Price 1993 is a sophisticated recent exposition of the theory.
[17] For more on the family resemblance view see Skyrms 1984a: 254.

like it, then it must be implying that if determinism is true then *every* action retroactively influences every past state of the world that was a determining cause of it, i.e. every non-hyperbolic section of its past light cone. It must be implying that human actions everywhere and always, and in contrast with all non-actions, are the causes as well as the effects of each of their own determining causes. Any theory that implies this seems to me not to be about *causation* at all.

Of course it is dogmatic to lay down just how incredible a theory of causality would have to be before we should stop calling it one. If you think that a theory of causation properly so-called *can* have consequences as amazing as this then *I* have no further objections, though others do.[18]

But anyway, no further objections are necessary. If the agency theory is true then EDT and CDT will coincide in their recommendations not only here but *everywhere*. A contemplated option will, from the deliberator's point of view, have an evidential bearing on any state of the world that exactly matches its causal bearing on that state. Therefore from that perspective the *V*-scale and the *U*-scale coincide. This rules out (a)- or (b)-type causalist arguments, as outlined at the start of Chapter 4. Moreover it is a short cut to my overall conclusion that the agent *need* make no use of specifically causal knowledge in practical deliberation. On the contrary, he need only care about the *evidential* bearing of his options upon whatever states of the world are of interest to him. Talk of their causal bearing will in that case only restate, in needlessly metaphysical terms, something that he already knew.

5.3.2   *The payment mechanism*

The causalist might say that CDT recommends O_2 in *Betting on the Past* because Alice's choice between O_1 and O_2 is causally relevant to what really matters to her: the money. After all, what causally determines what Bob pays her is *Bob's belief* about whether P is true or false. And what causes Bob to get this belief is what she does. The effect of her choosing O_1 is that Bob believes $\neg P$, so that she loses one dollar. The effect of her choosing O_2 is that Bob believes P, so she makes one dollar. So CDT recommends O_2.

---

[18] For further criticism of the Menzies/Price version of the agency theory, see Woodward 2003: 123–7. One further difficulty for the Menzies/Price theory is that it appears to imply an incredible form of causality even in the following case: Newcomb's problem, but the predictor has written a very large number (say, $10^{21} + 3$) on the top of the opaque box; and we know that the predictor writes a prime number if he thinks that you will one-box, but a composite number if he thinks that you will two-box. Then one-boxing is evidence for you, the agent, that this number is prime. So the Menzies/Price theory has the incredible consequence that you should therefore think that one-boxing *causes* it to be prime. But pursuing this matter involves relaxing the idealization (b2)(ii) from section 1.1, that you know all mathematical truths, so I will leave the matter there.

The simplest way around this is to suppose that the pay-offs in Tables 5.1 and 5.2 represent not moneys that Bob would give her but values that she *directly* attaches to the possible outcomes. For some reason she greatly prefers being a hand-raiser if P is true and not being one if P is false. This is indeed an unusual pattern of interests. But still the case has none of the science-fictional character of the standard Newcomb case, which traded on outlandish *beliefs*.

You might object that this pattern of interests is not just unusual but impossible, since you cannot aim at a state of which the only evidence is whichever of your own options the aim itself motivates. But this is an implausible restriction. It is plausible that such desires are needed to explain the writing of wills. And you might, like Magwitch, desire that your protegé do well in life and so arrange for his education, even though you are in permanent exile and expect never to learn of his success.[19]

A second way around the objection is to modify the example so that Bob can have evidence of P independently of Alice's actual choice. Suppose that Bob makes EEG recordings of Alice's cerebral activity by means of electrodes attached to her scalp; and let P be the proposition that her pre-decision cerebral activity together with L^* determines that she lowers her hand, where L^* now entails some favoured neurological theory. Libet's famous experiment showed that these recordings can reliably indicate the direction of her conscious decision 350–400 milliseconds before it occurs.[20] So Bob could use this evidence of Alice's state of mind just before her final decision as direct and perhaps decisive evidence of a P whose obtaining remains causally independent of anything that she does. All of the arguments against CDT would then go through as before. The only reason that I didn't choose such a P in the initial exposition was to make it especially *vivid* that P is causally independent of what Alice now does.

5.3.3 Backwards counterfactual dependence

Turning now to *Betting on the Laws*: it is perhaps not so obvious as I hoped to make it seem at section 5.2.1 that *anything* worth calling *Causal* Decision

---

[19] This example resembles one of Parfit's (1984: 495). Note that the present claim is *not* the one that Parfit discusses, that your children's success or failure is, though unknowable to you, still in some objective sense good or bad for you, although this may also be true. The present claim is only that their successes and failures would constitute the fulfilment or frustration of desires that you really have.

[20] By itself of course this does not show that Alice lacks free will, but only that some correlation obtains between her will and some past state of the world, which is hardly surprising. I need hardly say that I am skating over voluminous philosophical discussion of the Libet experiment. For more on it see e.g. Wegner 2002: 52–6. For its application to Newcomb cases see Slezak 2013: 16–17.

Theory recommends that Alice deny L^*. My own argument for the version discussed here went via a formulation of CDT in terms of conditional chances. But maybe other formulations do *not* support this claim and so evade the argument at section 5.2.

I have in mind versions of Causal Decision Theory formulated in terms of a counterfactual operator $>$, where $P > Q$ means something like this: If it were to be the case that P then it would be the case that Q. Then if $O \otimes P$ is a rich partition, O the set of options and P some partition of S into events, we can define:

(5.31)   *Counterfactual CDT*: A rational option $O \in O$ maximizes coun-
         terfactual utility, defined as: $U^*(O) = \Sigma_P \in {}_P Cr (O > P) V (O \wedge P)$.

Everything now turns on how we understand the counterfactual $>$. Before getting into that, note that it is at least intuitive that Counterfactual CDT gives the two-boxing recommendation in the standard Newcomb problem. For intuitively, what the predictor predicted is counterfactually independent of what you do: if e.g. you *were* to take both boxes, then his prediction *would* be the same as if you were to take just one box.

The standard approach to counterfactuals is the possible-worlds seman-tics: *A > C is true if and only if C is true at all the closest A-worlds to @.* But what does 'closest' mean? On the most influential reading,[21] and on any plausible modification of it, the whole manoeuvre is pointless, since Counterfactual CDT recommends O_2 in *Betting on the Laws*. Roughly speaking, that is because on this reading, the past remains counterfactually independent of the present: so if O_1 is a winning bet in *Betting on the Laws* then O_2 would also be a winning bet, in the '$>$' sense of 'would'; and if O_2 is a losing bet then O_1 would also be a losing bet in that sense; so it is still CDT-optimal to choose O_2.[22]

Still, there might be alternative interpretations of 'closest', and hence of '$>$', that evade this result. At least two theories of counterfactuals seem to yield operators of this sort: Bennett's 'exploding difference' proposal and Loewer's statistical mechanical (SM) proposal.[23]

On Bennett's account the closest A-worlds to a deterministic @ follow @'s laws everywhere and differ from it at all pre-A times, but only in *insignificant respects that could never matter to anyone*. It is when A comes true, and only then, that these worlds start visibly to diverge from @. On

---

[21] Lewis 1979a.   [22] For details of the argument see Ahmed 2013: 294–6.
[23] Bennett 2003: 217–8; Loewer 2007: 316 ff. Cantwell's theory exploits indicative rather counterfactual conditionals: for discussion of that approach see nn. 8 and 9 in this chapter.

Loewer's quite similar account, the closest A-worlds match @ at the time of A in all macroscopic respects excepting possibly A itself, and also match @'s laws, where these are stipulated to include the 'Past Hypothesis', a statement specifying the low-entropy macro state of the universe at one boundary.[24] It can then be shown that the Loewer-closest P-worlds will match @ over all pre-A matters of macroscopic fact that leave macroscopic traces at the time of A, but *not* in all other respects.

Because of the emphasis that each of these counterfactuals (call them $>_B$ and $>_{SM}$) places on the preservation of law, L^* is counterfactually independent of what Alice does. So we have:

(5.32) $\quad Cr\,(O_1 >_B L^*) = Cr\,(O_1 >_{SM} L^*) = Cr\,(L^*)$
(5.33) $\quad Cr\,(O_2 >_B \neg L^*) = Cr\,(O_2 >_{SM} \neg L^*) = Cr\,(\neg L^*)$

Since $Cr\,(L^*) \gg 0.5$, it follows from (5.31)–(5.33) and Table 5.3 that on both interpretations of '$>$', Counterfactual CDT agrees with EDT and with intuition that O_1 is uniquely rationally optimal.

But (i) *ad* Bennett: there may not *be* any Bennett-closest A-worlds for certain choices of A that are surely fit subjects of decision theory. Thus if L^* is true and Alice realizes O_1, then any legal O_2-worlds differ from @ over P. If Alice realizes O_2, then any legal O_1-worlds differ from @ over P. Either way, there is an A such that any legal A-world differs from @ in some respect that *does* matter to somebody, viz Alice. So for at least one of these choices of A, no legal exploding-difference A-world exists.

(ii) *Ad* Loewer: no decision theory that recommends maximizing counterfactual utility relative to $>_{SM}$ should call itself *causal*. Notice that since the Loewer-closest A-worlds differ from @ over the past, any decision theory that incorporates this measure of closeness forces somebody who is contemplating A to compare possible worlds that differ even over their pasts. Hence to advocate this reading of Counterfactual CDT is to advocate making choices as though *causally unreachable aspects of the past* turned upon one's present choice. If *that* doesn't stop it from counting as a Causal Decision Theory then I don't know what stops anything from counting as a Causal Decision Theory.

Loewer agrees that '[i]t is a consequence of this account that the probability of the past micro state is correlated with present alternative decisions'.

> But [he continues] this does not mean that a person can affect the past in the sense of having control over past micro events. Control by decision requires that there be a probabilistic correlation between the event of deciding that p be so and p being so and one's knowing (or believing with reason) that the

---

[24] Loewer 2007: 300.

correlation obtains. But it is immensely implausible that there is any past micro state *m* that fulfills the first part of this condition let alone both parts. So while it is true on the account that if Nixon had pushed the button the probability that the past would have been different in some micro respects is 1. But since we have no idea what these micro respects are we have no control over past micro conditions.[25]

But first: yes, there may be *a* sense in which you 'control' something only if you know that you do. But we can still say that something *turns on* your decision whether or not you know exactly what the correlation is. I might know, and it might matter to me, that the state of the world in a million years' time somehow turns on what I do now, though in respects that are completely unfathomable to me now. This might be a way of reconciling myself with mortality. But $>_{SM}$-dependence implies that in *this* sense the state of the world a million years ago *also* turns on what I do now. The point is not that this makes the associated decision theory false but only that it makes it non-causal.

Second, Loewer claims that 'it is immensely implausible that there is any past micro state that fulfills the first part of this condition let alone both parts'. But there *is* a *partition over past micro states* that fulfils both, namely $\{P, \neg P\}$ where *P* is as described in *Betting on the Past*. It is surely possible for Alice to have the reasonable belief that determinism is true and that she is faced with a choice as to whether to bet on *P*. In that case, she also reasonably believes that the past event described by *P* is correlated with her present decision.

So even in Loewer's sense, which demands reasonable belief in the correlation, his account entails that Alice has control over the past event described by *P*. This is no reason to reject the statistical mechanical approach to decision theory. But it is all the more reason to distinguish it from the *causal* theories that these examples are supposed to refute. For although Alice does have $>_{SM}$-control over whether *P*, she does not have causal control over whether *P*.

What may be a reason to reject the SM approach is that when viewed in this light it starts to look unmotivated. One can at least see the point of a decision theory that evaluates options on the basis of their causal effects. Similarly one can see the point of a theory that evaluates them on the basis of their news value. But to evaluate them on the basis of neither of these but rather on the basis of $>_{SM}$ is arbitrary.

In particular, when evaluating a decision-theoretically relevant counterfactual of the form $O >_{SM} S$ we are supposed to hold fixed all macro

[25] Loewer 2007: 318.

features of the actual world that are contemporary with the action that O describes. So one is supposed to evaluate an option by comparing it with alternatives in worlds that match the contemporary macro state of @, for instance over whether there is now 1 million dollars in the opaque box, even if one takes there to be a probabilistic correlation between one's actions and other contemporary features of the macro state.

What is the justification for this? The only obvious one is that those features are *causally* independent of whether the agent now realizes O. But in that case we should also hold fixed the past micro state of the world, as this too is causally independent of whether the agent now realizes O. But that is just what we do *not* hold fixed in the evaluation of $>_{SM}$. This justification of 'statistical mechanical decision theory' is therefore unstable. Consistently applied, it supports not the SM theory itself but the *Causal Decision Theory* against which *Betting on the Past* and *Betting on the Laws* were directed all along.[26]

5.4 The openness of the past

Finally let me briefly say how these examples cast new light on the contrast between the openness of the future and the fixity of the past. More precisely, they undermine this contrast on one interpretation of it.

The interpretation that I have in mind is an asymmetry in what deliberating agents take themselves to be deliberating *between*. We normally take the options that are open to us, the possible worlds from which we are choosing which to actualize, to have *different* futures but a *common*

---

[26] Loewer (2007: 323) does mention another possible justification: 'People whose degrees of belief approximate the statistical mechanical probability distribution are objectively more likely to succeed in satisfying their desires (assuming they are otherwise rational) than people whose degrees of belief diverge from this distribution'. Here 'the statistical mechanical probability distribution' is an objective distribution PROB that is uniform over those possible initial micro conditions of the universe compatible with the proposition *PH* that specifies its initial macro condition, including its low entropy at that time. And the 'otherwise rational' people whose degrees of belief approximate PROB are those (a) for whom $Cr(O >_{SM} S)$ is close to PROB $(S \mid O \land PH \land L \land M(t))$, where L specifies the true, deterministic dynamical laws and $M(t)$ specifies the macro state of the rest of the universe at the time that O describes; (b) who maximize counterfactual utility relative to $>_{SM}$. Now if Loewer means that these people will actually get what they want more *frequently* than people who act in alternative ways, then the claim is simply false. People who satisfy (a) and (b) will inevitably two-box in Newcomb's problem and so will actually only ever get 1,000 dollars from it, whereas many 'irrational' people will frequently get 1 million dollars. If Loewer means that his 'rational' people *would* do better than if they themselves *were* to act otherwise, where 'would' is interpreted as the everyday counterfactual, then the claim is again false: in *Betting on the Past* Alice would in that sense get more if she were to realize O_1 than if she were to realize O_2, but Loewer's 'rational' people would prefer O_2. Finally, if he means that his 'rational' people would do better than if they themselves were to act otherwise, where 'would' is interpreted as his own $>_{SM}$, then the claim is true but is no justification for counterfactual utility maximization relative to $>_{SM}$. Analogous 'justifications' could equally be given for U-maximization and for standard U^*-maximization.

past. This metaphysical assumption is visible in the common metaphor for moments of decision as *forks* in a road. At a fork you decide between paths that diverge at *that* point, not between ones that were separate *all along*. Call this the *asymmetry of deliberative openness*.

What justifies this asymmetry in our thought about decisions? One obvious source of support is the corresponding *causal* asymmetry. Events and states of affairs are generally causally *dependent* upon some other events in their *past* but causally *independent* of all those in their *future*. Setting aside (a) the agency theory of causation as discussed in section 5.3.1, and (b) troublesome quantum cases as discussed in Chapter 6, this applies in particular to human actions. Nothing that I *now* do makes any causal difference to the past. But anything that I now do makes a causal difference to the future. And nothing in the present discussion casts any doubt on this *asymmetry of causal openness*.

Causal Decision Theory encourages the step from the asymmetry of causal openness to the asymmetry of deliberative openness. One way to put the idea behind CDT is that when choosing between options, agents should compare possible worlds that differ over those options but *not* over anything causally independent of the options. It follows from the asymmetry of causal openness that they should compare worlds that differ over the future but not over the past. This rationalizes the metaphysical picture that constitutes the asymmetry of deliberative openness.[27]

But Alice's problems show that, at least if we are confident that determinism holds, CDT errs on just this point. For instance, if in *Betting on the Past* Alice compares worlds that realize options O_1 and O_2 but which agree on everything that is *causally* independent of her choice between them, she will be comparing worlds that agree over P. In any *such* comparison, the O_1-worlds will always do better. Instead and as I have argued, she ought to compare realizing O_1 when P is false with realizing O_2 when, and because, P is *true*. This comparison gives the correct recommendation, that she should choose O_2. An analogous point applies to *Betting on the Laws* with Q, as defined at section 5.2.1, in place of P.

And that rationalizes a temporally *symmetric* picture of deliberative openness. Deliberating agents should take their choice to be between worlds that differ over the past as well as over the future. In particular, they differ over the *effects* of the present choice but *also* over its unknown causes. Typically these past differences will be microphysical differences that don't matter to anyone. But in *Betting on the Past* they matter to Alice.

---

[27] For instance, Mellor (1987) advocates this connection between causal and deliberative openness.

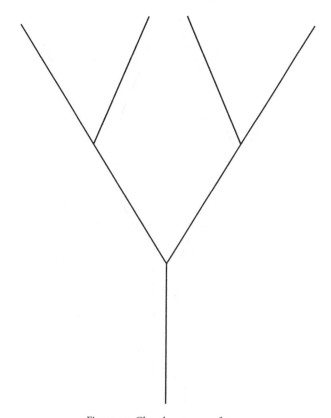

Figure 5.1 Closed past, open future

On this new picture, which arises naturally from EDT but also from the statistical mechanical theory as discussed at section 5.3.3(ii), it is misleading to think of decisions as forks in a road. Rather, we should think of them as choices between roads that were separate *all along*. For instance, in *Betting on the Past*, Alice should take herself to be choosing between actualizing a world at which *P* was *all along* true and one at which it was *all along* false. More generally, we should think of decisions not as the nodes in Figure 5.1 but as the branching points in Figure 5.2.

In Figure 5.2, each line represents a possible evolution of the state of the world. That adjacent lines on a branch are parallel at a time represents that they are macroscopically indistinguishable at that time. So at each branching point the agent is deciding between possibilities that were different *all along* but which only became *visibly* different at the point of

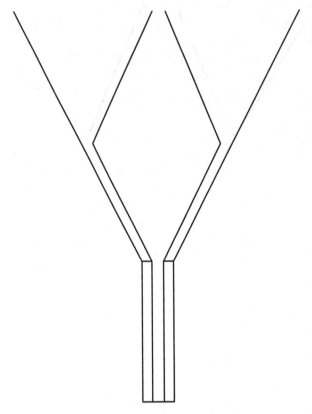

Figure 5.2 Open past, open future

decision. This corresponds to the fact that in *Betting on the Past*, Alice's
choice is the first *visible* difference between otherwise similar worlds that
disagree over *P*. More generally, it reflects Loewer's observation that human
decision-makers possess a biological structure on which 'very small differences in the brain get magnified into differences in bodily movements
and these, in some cases... get magnified into vast differences in the
world'.[28]

I should emphasize that this second picture is entirely compatible with
the temporal asymmetry of *causal* openness. It is not making any claims
about causal dependence, and in particular it does not take the agent to

---

[28] Loewer 2007: 317.

have any causal influence on the past. It simply represents the worlds that the agent should take herself to be choosing between when deliberating. And as Figure 5.2 shows, these worlds differ over the causally inaccessible past as well as over the causally accessible future. What the examples of this chapter vindicate, and what this new metaphysical picture expresses, is therefore precisely that feature of EDT that distinguishes it from causalism.

CHAPTER 6

Quantum-mechanical cases

This chapter describes a family of practically feasible decision situations in which EDT and CDT give conflicting recommendations and bases two criticisms of CDT upon them. First and as with *Betting on the Past*, I argue that CDT gives what are intuitively and also observably wrong recommendations in these cases (if it gives any at all). Second, I argue that CDT wrongly makes the practical question what one should *do* in these cases turn on possibly irresoluble metaphysical speculations about their causal structure.

The feasibility of these cases follows from experimentally observed quantum statistics. In fact they all rely upon a device that essentially involves quantum phenomena. But this doesn't mean that their outcomes are themselves only observable at a quantum scale. Nor does the argument rely on any specifically quantum-mechanical reasoning. In fact it is possible to explain the decision problem in terms that are completely free of those technicalities, as follows.

6.1 The device

At the heart of all these problems is a device having three components: a source S and two receivers A and B. The receivers stand on either side of the source and are so separated from one another, by very large distances, impenetrable barriers etc., that there is no possibility of causal commerce between them. At least, we have as good non-statistical evidence that *they* are causally isolated as we ever have that *any* two systems are causally isolated. Each receiver features a display and a switch with three settings labelled 1, 2 and 3. We can independently put either switch in any of these settings.

After setting the switches we activate S. It emits two signals, each receiver picking up one. The display on each receiver then shows one of two readings: 'y' or 'n'. That represents one 'run' of the device. We record the run by noting down the setting of each receiver and the reading on each display. For instance, '12yn' says that A was set to 1, B was set to 2, the display on A was y and the display on B was n. Similarly, '33nn' says that both receivers were set to 3 and both displayed 'n'.

We repeatedly run the device with the receivers being set at random. Each possible setting of the receivers occurs with the same frequency. Repeated runs reveal the following statistical facts:

(6.1)   *Whenever* the switches on A and B are on the *same* setting (i.e. both on 1, both on 2 or both on 3) the devices display the same reading: either both say 'y' or both say 'n'. So we sometimes get runs like this: '11yy', '22yy'. But we *never* get runs like this: '22yn', '33ny'.

(6.2)   When the switches on A and B are on *different* settings (e.g. A on 1, B on 3), the devices display the same reading about 25 per cent of the time. So we get runs like '12yn' and '23ny' about three times as often as we get runs like '12yy' and '13nn'.

That is all we need to know about the device. Quantum theory predicts that an apparatus could work in this way and conform to these statistics.[1]

The only other assumption is that the two receivers are as causally independent as they appear to be. In particular, I will suppose until section 6.5.1 that the setting on receiver A, and its reading, on any particular run of the device, are both causally irrelevant to the reading on receiver B on that run. I'll also assume that neither the setting nor the reading on receiver B makes any difference to the reading on receiver A. More briefly, I'll assume that any agent in the following decision situations is certain of this:

(6.3)   **Causal independence**: whatever happens at either receiver is causally independent of anything that happens at the other.

I will also make an additional assumption, not about the device but only about any agent who uses it. This is the *F-C connection*: the statistics (6.1) and (6.2) control your conditional credences concerning the outcome of any future run of the device. ('F-C' is for 'frequency-credence'.) For instance, your credence that the readings on both receivers will be the same on the next run, given that both receivers are at the same setting, is 1. Your credence that the readings will both be the same on the next run, given that the receivers are at different settings, is 0.25, etc. Again this assumption is hardly outlandish. It is quite natural that one's credences concerning *any* repeatable phenomenon should at least be sensitive to the observed statistics, especially when these are as well attested as (6.1) and (6.2) in fact are.

---

[1] For details see Mermin 1981: 407–8.

Table 6.1 D_i (z)

	yy	yn	ny	nn
*ii*hom	$1-z$	$-z$	$-z$	$1-z$
Q	0	0	0	0

6.2   Identical settings

Suppose that we are about to run the device. I offer you the following bet. I will set both receivers to the *same setting i* on the next run, where I have chosen $i = 1$, 2 or 3. If they then display the *same reading* you win one dollar. If they display *different* readings then you win nothing. To accept the bet you must first pay me z, for some z such that $0 \le z \le 1$. Alternatively you can reject the bet, in which case you will get nothing. Let us call this *the decision situation D_i (z)*.

If, for reasons that will soon be clear, we label the first option '*ii*hom', and if we label the second option Q ('quit'), and if we assume as usual that value is dollar value, we can represent D_i (z) by means of Table 6.1. In it, the two letters at the head of each column represent the readings on receiver A and receiver B respectively. What do EDT and CDT advise?

First consider EDT. Fact (6.1) and the F-C connection entail that the relevant conditional credences are as follows:

(6.4) Cr (yy $\lor$ nn$|ii$hom) $= 1$

(6.5) Cr (yn $\lor$ ny$|ii$hom) $= 0$

It follows that the *V*-scores of the options in D_i (z) are:

(6.6) V (*ii*hom) $= 1 - z$

(6.7) V (Q) $= 0$

Hence for any $i = 1$, 2, 3 and z such that $0 \le z \le 1$, EDT will at least endorse playing *ii*hom in D_i (z). If in addition $z < 1$ it will strictly prefer *ii*hom to Q. So it always endorses and sometimes requires your betting that the receivers give the same reading when at the same setting.

What about CDT? The relevant K-partition will have elements of the form:

(6.8) K_x: Ch (yy $\lor$ nn $|$ *ii*hom) $= x \land Ch$ (yn $\lor$ ny $|$ *ii*hom) $= 1 - x$.

So if Cr (K_1) < 1 there is some $z^* < 1$ such that U (Q) $> U$ (*ii*hom) in D_i (z) for any $z \ge z^*$.[2] In English: unless you think that setting the receivers

---

[2] In D_i (z), $U_z =_{def.} U$ (*ii*hom) $= \int_0^1 ((1 - z)x - z(1 - x))f(x)dx = \int_0^1 (x - z)f(x)dx = \int_0^1 xf(x)dx - z$, where $f = F'$ and $F(x) = Cr$ ($\cup_{0 \le j \le x} K_j$). So if Cr (K_1) < 1 then $U_z < 1 - z$ for all z. Since U_z is a

to the same setting *makes* it certain that they will give the same reading, CDT will advise you to walk away from this bet for some dollar fee z^* less than one dollar.

So if $Cr(K_1) < 1$ things look like this. You have the option of paying $\$z$ for a bet that pays one dollar if you win and zero if you lose. You win if (6.1) is reliable. So the evidentialist will pay any fee short of one dollar to take this bet. But the causalist will decline the bet at any fee beyond some threshold $\$z^*$, where $z^* < 1$.

So if both are offered these bets for $\$z$ with $z \geq z^*$, the causalist will keep declining and keep winning nothing, and the evidentialist will keep accepting and keep winning $\$(1 - z)$. For instance, suppose we have $z^* = 0.8$. Then we can keep charging both parties 90 cents for a bet that pays one dollar iff both receivers give the same reading on the next run in which they are switched to the same setting. The evidentialist will always accept and the causalist will always decline, and the evidentialist will make 10 cents over the causalist every time. So here we have a decision problem where EDT and CDT disagree.

Or rather they disagree on a certain assumption, namely that for some i the factor $Cr(K_1) < 1$. If we drop this assumption then the causalist only declines the bet at $z \geq z^* = 1$, at which rate EDT will also endorse not betting, since the value of iihom is then certainly non-positive. So assuming that $0 \leq z \leq 1$, there is in that case no $D_i(z)$ on which EDT and CDT disagree.

But as I'll now argue, if we *do* drop that assumption then there is inevitably *another* quantum situation in which EDT and CDT disagree. Dropping the assumption implies that:

(6.9) $Cr(K_1) = 1$ i.e. $Cr(Ch(\text{yy} \vee \text{nn} \mid ii\text{hom}) = 1) = 1$.

So we may retain *this* throughout what follows.

6.3 Non-identical settings

In this next case, which I'll call **EPR** just for the association with quantum mechanics, you *must* take one of six *free* bets, as follows. You can choose any of three joint settings for each receiver: A on 1 and B on 2, A on 1 and B on 3, or A on 2 and B on 3. And for each setting you can bet either that the receivers will display the *same* reading on the next run, or that they will display a *different* reading on the next run. I'll label the six resulting options 12hom, 13hom, 23hom, 12het, 13het and 23het. For instance, '12het'

---

continuous function of z there is therefore some $z^* < 1$ s.t. for all $z \geq z^*$ $U_{z^*} < 0$. So CDT prefers Q to iihom in $D_i(z)$ for all $z \geq z^*$.

Table 6.2 *EPR*

	yy	yn	ny	nn
12hom	2	0	0	2
13hom	2	0	0	2
23hom	2	0	0	2
12het	0	1	1	0
13het	0	1	1	0
23het	0	1	1	0

represents the option that you set receiver A to setting 1, receiver B to setting 2, and bet that they will give *different* readings, either 'yn' or 'ny'.

The pay-offs now depend on whether you bet 'hom' or 'het'; that is, on whether you bet that the receivers will give the same or different readings on the next run. In particular, the pay-off to a winning 'hom' bet is twice the pay-off to a winning 'het' bet. Your values for the outcomes are as in Table 6.2. What does EDT recommend? Again, the F-C connection settles this straightforwardly. It and the statistical fact (6.2) entail that for $1 \leq i < j \leq 3$:

(6.10) $Cr\,(yy \vee nn|ij\text{hom}) = Cr\,(yy \vee nn|ij\text{het}) = 0.25$

(6.11) $Cr\,(yn \vee ny|ij\text{hom}) = Cr\,(yn \vee ny|ij\text{het}) = 0.75$

So:

(6.12) $V\,(ij\text{hom}) = 2(0.25) = 0.5$

(6.13) $V\,(ij\text{het}) = 1\,(0.75) = 0.75$

So EDT prefers every 'het' option to any 'hom' option. It says you should always bet that A and B give *different* readings on the next run.

What about CDT? We need to consider the possible chances of each possible pair of readings given each possible setting. The reasoning turns on four points.

(i) Clearly the choice between a 'het' and a 'hom' bet makes no difference to the reading on either receiver once we are given their settings. We could in any case impose this condition by brute force, by requiring that you choose between 'hom' and 'het' *after* the run is over but *before* you see the readings. So we can write:

(6.14) $Ch\,(yy|12) =_{\text{def.}} Ch\,(yy|12\text{het}) = Ch\,(yy|12\text{hom})$ etc.

(ii) Since the readings are by (6.3) *causally* independent of one another, the *chance* of either reading on either receiver is independent of the reading on the other receiver, even given the settings on both receivers. So if we write I_A, y_B etc. for the settings and the readings on each receiver, then we have e.g.:

(6.15) $Ch\,(yy|12) = Ch\,(y_\text{A}y_\text{B}|1_\text{A}2_\text{B}) = Ch\,(y_\text{A}|1_\text{A}2_\text{B})\,Ch\,(y_\text{B}|1_\text{A}2_\text{B})$

(6.16) $Ch\,(yy|11) = Ch\,(y_\text{A}y_\text{B}|1_\text{A}1_\text{B}) = Ch\,(y_\text{A}|1_\text{A}1_\text{B})\,Ch\,(y_\text{B}|1_\text{A}1_\text{B})$ etc.

As (6.16) illustrates, this point applies to chances conditional on any pair of settings, including those that the present decision problem does not associate with any pay-offs.

(iii) Again by (6.3), the reading on either receiver is causally independent of the *setting* on the other receiver. This means that the setting on either receiver is irrelevant to the chance of any particular reading on the other receiver. So we have e.g.:

(6.17) $Ch\,(y_\text{A}|1_\text{A}2_\text{B}) = Ch\,(y_\text{A}|1_\text{A})$

(6.18) $Ch\,(y_\text{B}|2_\text{A}2_\text{B}) = Ch\,(y_\text{B}|2_\text{B})$ etc.

Again and as (6.18) illustrates, this point applies to all pairs of settings, not only those to which *EPR* assigns a positive pay-off.[3]

(iv) The fourth point involves the family of decision problems $D_i\,(z)$. Recall that if EDT and CDT agree over all of those cases then (6.9) must be true. And if they do not, then we have already got a decision-theoretic case over which they disagree and to which both of the forthcoming arguments, at section 6.4 below, will apply. So we may assume (6.9). It follows that you know:

(6.19) $Ch\,(yy \vee nn|11) = 1$

(6.20) $Ch\,(yy \vee nn|22) = 1$

(6.21) $Ch\,(yy \vee nn|33) = 1$

Focusing on (6.19), we see that:

(6.22) $Ch\,(yy|11) + Ch\,(nn|11) = 1$

And so by (6.16) and its analogue for 'nn', and (6.17) and (6.18) and *their* analogues for n_A and n_B, we have e.g.:

---

[3] Note that points (i)–(iii) suffice to derive a Bell inequality for the *Ch* function. This does *not* contradict the predictions of quantum mechanics so long as the chances given by *Ch* do not reflect long-run relative frequencies. This is permitted if, for example, *Ch* represents single-case chances which vary from case to case. And that is what *Ch* *should* represent if conditional chance matters to *Causal* Decision Theory: for CDT is supposed to be sensitive to the tendency of a setting to causally promote this or that outcome. The situation here is similar to that in Newcomb's problem, where, even though there is a long-run correlation between one's choosing one box and this having been predicted, the latter is, on any occasion, conditionally *independent* of the former with respect to the appropriately causal chance function. This is consistent with the claim that chances control long-run frequencies if, as in the Newcomb case and as here, these one-off conditional chances vary from one occasion to the next.

(6.23) $Ch\,(y_A|_{1A})\,Ch\,(y_B|_{1B}) + Ch\,(n_A|_{1A})\,Ch\,(n_B|_{1B}) = 1$ etc.

And corresponding results follow from (6.20) and (6.21). Finally, from (6.23) and its analogues for $2_A, 2_B, 3_A, 3_B$ we get:[4]

(6.24) $Ch\,(y_A|_{1A}) = Ch\,(y_B|_{1B}) \in \{0, 1\}$
(6.25) $Ch\,(y_A|_{2A}) = Ch\,(y_B|_{2B}) \in \{0, 1\}$
(6.26) $Ch\,(y_A|_{3A}) = Ch\,(y_B|_{3B}) \in \{0, 1\}$

This gives eight possibilities for the values of these conditional chances, depending on which ones take the value 1 and which take the value 0.

Here is a more informal version of the argument for point (iv). Suppose you have a sequence of boxes labelled 1, 2, ... Each box contains a very large number of tickets stamped 'YES' or 'NO'. On the first trial I make two draws from box 1, on the second trial two from box 2, and so on. I know that on any particular trial the first draw *has no effect* on the second. Now suppose that on every trial I find either that *both* tickets are stamped 'YES' or that *both* are stamped 'NO'. This leads me to conclude that the chance of my drawing either two 'YES' tickets or two 'NO' tickets on any particular trial is 1. The argument is then that if this is so, then it must also be true that on any particular trial, the chance that the *first* draw is of a 'YES' ticket must be 1 or 0, and the chance that it is a 'NO' ticket must be 0 or 1. Similarly, the chance that the *second* draw in that trial is of a 'YES' ticket equals the corresponding chance for the first draw, i.e. 1 or 0. Otherwise there would have to be a non-zero chance of drawing two *differently* stamped tickets on any particular trial.

So the conditional chance of each reading (yy, yn etc.) on each option (12hom, 13het etc.) is either 1 or 0. This is determined by which of the eight possibilities just outlined obtains. For instance, suppose that the following situation obtains:

(6.27) $Ch\,(y_A|_{1A}) = Ch\,(y_B|_{1B}) = 1$
(6.28) $Ch\,(y_A|_{2A}) = Ch\,(y_B|_{2B}) = 0$
(6.29) $Ch\,(y_A|_{3A}) = Ch\,(y_B|_{3B}) = 1$

Then by (6.14), (6.15), (6.17), (6.27) and (6.28) it follows that:

(6.30) $Ch\,(yy|12het) = 0$

---

[4] Proposition (6.23) has the form $xy + (1 - x)(1 - y) = 1$. So for $0 \le x, y \le 1$ the only solutions are $x = y = 0$ and $x = y = 1$. I am here assuming that we are not invoking negative probabilities, as some quantum theorists have suggested. See e.g. Muckenheim 1982.

Table 6.3 *EPR* with elements of the K-partition

	111	110	101	100	011	010	001	000
12hom	2	2	0	0	0	0	2	2
13hom	2	0	2	0	0	2	0	2
23hom	2	0	0	2	2	0	0	2
12het	0	0	1	1	1	1	0	0
13het	0	1	0	1	1	0	1	0
23het	0	1	1	0	0	1	1	0

More generally, any specification of the conditional chances in (6.24)–(6.26), together with a specification of your option, completely determines the reading on the receivers and hence your pay-off.

It follows that we can rewrite the decision problem in Table 6.2 in terms of a K-partition for whose elements I'll use the following three-digit code: 'abc', for $a, b, c \in \{0, 1\}$, means:

(6.31) $abc \equiv_{\text{def.}} Ch(y_A|1_A) = Ch(y_B|1_B) = a \wedge Ch(y_A|2_A) = Ch(y_B|2_B) = b \wedge Ch(y_A|3_A) = Ch(y_B|3_B) = c$

For instance, '101' corresponds to the possible distribution stated at (6.27)–(6.29). We can now represent the pay-offs in *EPR* like this in Table 6.3. Calculation of the U-score for each option is now straightforward. By (2.29) it follows from Table 6.3 that the U-scores for the three 'hom' options are:

(6.32) $U(12\text{hom}) = 2\,(Cr(111) + Cr(110) + Cr(001) + Cr(000))$
(6.33) $U(13\text{hom}) = 2\,(Cr(111) + Cr(101) + Cr(010) + Cr(000))$
(6.34) $U(23\text{hom}) = 2\,(Cr(111) + Cr(100) + Cr(011) + Cr(000))$

And the U-scores for the three 'het' options are:

(6.35) $U(12\text{het}) = Cr(101) + Cr(100) + Cr(011) + Cr(010)$
(6.36) $U(13\text{het}) = Cr(110) + Cr(100) + Cr(011) + Cr(001)$
(6.37) $U(23\text{het}) = Cr(110) + Cr(101) + Cr(010) + Cr(001)$

Finally, the following argument shows that if (6.32)–(6.37) are true then there must be some 'hom' option that CDT endorses. Suppose on the contrary that CDT (in agreement with EDT) prefers any 'het' option to every 'hom' option. Then all of the following must be true:

(6.38) $U(12\text{het}) > U(12\text{hom})$
(6.39) $U(13\text{het}) > U(13\text{hom})$
(6.40) $U(23\text{het}) > U(23\text{hom})$

Table 6.4 Simplified *EPR*

	yy	yn	ny	nn
12hom	2	0	0	2
13hom	0	0	0	0
23hom	0	0	0	0
12het	0	1	1	0
13het	0	0	0	0
23het	0	0	0	0

Substituting (6.32)–(6.37) into (6.38)–(6.40) and adding the three resulting inequalities gives:

$$(6.41) \quad 2\,(Cr\,(110) + Cr\,(101) + Cr\,(100) + Cr\,(011) + Cr\,(010) + Cr\,(001)) > 6\,(Cr\,(111) + Cr\,(000)) + 2\,(Cr\,(110) + Cr\,(101) + Cr\,(100) + Cr\,(011) + Cr\,(010) + Cr\,(001))$$

So:

$$(6.42) \quad 0 > 6\,(Cr\,(111) + Cr\,(000))$$

But (6.42) is false, since the credences on its right hand side are both *at least* zero. And so the supposition that entails (6.38)–(6.40) must be false. There must be *some* 'hom' option that gets at least as high a U-score as its corresponding 'het' option:

$$(6.43) \quad \text{For some } i, j \text{ s.t. } 1 \leq i < j \leq 3, \ U\,(ij\text{hom}) \geq U\,(ij\text{het})$$

So (6.12), (6.13) and (6.43) imply that there is *some* pair of options ijhom and ijhet over whose relative ranking EDT and CDT disagree.

In particular, let this be the pair 12hom and 12het. Then $V\,(12\text{het}) > V\,(12\text{hom})$ and $U\,(12\text{hom}) \geq U\,(12\text{het})$. Now consider the same decision situation as before but with pay-offs that make irrelevant all of the options except for these two (Table 6.4). Plainly nothing about the new pay-offs to the *other* options makes a difference to the V-scores and the U-scores of 12hom and 12het, over whose relative merit EDT and CDT will still disagree. It's also clear that *both* EDT *and* CDT take 12hom and 12het to be at least as good as any other option in this case. So in the case that Table 6.4 describes, EDT and CDT make different recommendations. EDT recommends only 12het, which gets a V-score 0.75 over 0.5 for 12hom. CDT endorses 12hom, which is getting some unknown U-score that is no less than $U\,(12\text{het})$.

Since the device that I have described is certainly feasible, we therefore have a realistic case over which EDT and CDT genuinely clash. And it lacks all the psychological clutter of 'tickles' and other forms of self-knowledge that so gummed up many previous attempts to come up with realistic cruces. It is only necessary that you be certain of (6.3), the claim that nothing that happens at either receiver has any causal bearing on what happens at the other. As I have already suggested we can arrange things to make this plausible to common sense, for instance by placing arbitrarily thick barriers or arbitrarily great distances between A and B. This does nothing to disturb the statistical facts (6.1) and (6.2) that are the only other prerequisites of the case.[5]

6.4 QM versus CDT

This disagreement between EDT and CDT does not by itself show which one is wrong. But it does make particularly vivid what is involved in following CDT and so gives a certain realism to an argument against CDT that I have already mentioned: 'Why Ain'cha Rich?'

Consider *Simplified EPR* in Table 6.4. EDT advises you to take 12het and CDT endorses 12hom. Which is right? When adjudicating between theories, philosophers often resort to thought experiments. That is now unnecessary because we *already know* the results of a *real* experiment.

Suppose that an evidentialist and a causalist with a common under-standing of this problem[6] took repeated bets on it in accordance with their favoured decision theory. So the evidentialist would repeatedly take 12het and the causalist 12hom. At least, this is so if the causalist's credences conform to (6.9). If he does not accept (6.9) then the same argument is available with a different decision problem, namely $D_i (z)$ for some $z < 1$.

Then we already know from fact (6.2) that the evidentialist wins one dollar in three out of every four runs and the causalist wins two dollars in one out of every four runs. So we know in advance that the evidentialist

---

[5] It's worth contrasting this construction with two other attempts (the only ones known to me) to exploit quantum violations of the Bell inequalities in order to make EDT and CDT disagree. Berkovitz's example (1995) assumes that you are simply unaware of the theoretical results that make this case so mysterious. As a result, you approach the problem with a demonstrably false assumption. In that situation it is not surprising, or especially damaging to causalism, that CDT gives you advice that loses money to a bookie that knows about these things. Cavalcanti's argument (2010), which invokes the CHSH arrangement (Clauser et al. 1969), appears to mischaracterize CDT itself. His case depends crucially on there being *two* agents, one at each wing of the experiment. But his calculation of the *U*-score of any option available to *one* of these agents treats both agents' choices as actions, i.e. ignores their evidential bearing on anything other than their effects. (A formal symptom of this is the symmetrical treatment of the terms 'A_R' and 'B_G' in his equation (16) (Cavalcanti 2010: 585).) But this is a mistake: from the point of view of either experimenter the other agent's choice – which is not up to *her* – itself partly characterizes the state of nature, and her credence should reflect this. Cavalcanti's reasoning that the causalist must bet against quantum mechanics in these scenarios (ibid.: 585–6) is therefore invalid.

[6] In particular, they both believe (6.1)–(6.3) and the F-C connection applies to both of them.

will make \$1.50 out of this game for every dollar that the causalist makes. Who do you think has the better strategy?

Let me put it as strongly as this. Suppose you and I take turns to choose an option from Table 6.4. On your turns, I pay you what you win. On my turns, you pay me what I win. So if I follow EDT and you follow CDT then I will on average win one dollar from you every four runs. I hereby publicly challenge any defender of CDT who accepts (6.3) to play this game against me. In fact he needn't fully accept (6.3): as we'll see at section 6.5.2 below, it is possible to adapt the game so as to make money out of any defender of CDT who gives *some* credence to (6.3).[7] I could finance the device by means of a loan: any banker to whom I offered the foregoing argument would certainly make me an advance on generous terms. Unfortunately I expect that Lewis, Pearl et al. would, if faced with this situation, stop being causalists long before they stopped being solvent.

The second objection to CDT is not that it gives *bad* advice, but that *what* advice it gives depends not on the observable statistics but on whether (6.3) is acceptable, i.e. whether or not there *is* any causal influence between the receivers. As we'll see at section 6.5.1, some philosophical analyses of causation agree with intuition in counting (6.3) true, and others reckon it false. For now, what matters is not which analysis is correct, but that *what* advice CDT gives *depends* on settling this philosophical question.

For it should seem strange that the answer to a practical question ('Which bet?') should turn on abstruse theoretical matters. After all, nothing about the theoretical situation has any impact upon the facts that will actually settle your pay-offs. We know in advance what these are. We know in advance that whether or not (6.3) is true, the expected return to 12het in Table 6.4 will exceed that to 12hom by 50 per cent.

This complaint against CDT goes to the heart of what distinguishes causalism from evidentialism. Causalism makes a practical question about what to do depend upon recondite metaphysical issues on which maybe *nothing* observable turns. That involves it in a complete misconception of what practical reasoning is and why it matters. To give non-trivially different practical advice in practically indistinguishable situations is to fail to understand that you are giving *practical* advice.

This aspect of CDT is absent from the other cases that distinguish it from EDT. In the standard Newcomb case the causal structure of the situation is clear because stipulated: there simply *is* no causal route from your decision

---

7 For evidence that Lewis was at one time such a person see Butterfield 1992b: 41 – although as that paper argues, it is another question whether Lewis's own (1973a) theory of causation is compatible with (6.3).

to the prediction.[8] The same is true in cases not involving dominance, such as Egan–Gibbard cases (section 3.1) or remedial cases (section 4.1). So although it is (in my view) always true that the statistical facts are enough by themselves for practical purposes, it is only in the quantum cases here discussed that *they* are clear but the underlying causal structure is not. *These* cases reveal that CDT is implausibly sensitive to background theorizing about a device whose actual outputs are already known to everyone.

6.5 Objections

The defender of CDT might object that on some theories of causality CDT makes the *same* recommendation as EDT in *EPR*; also that the argument does not apply if you are uncertain of (6.3).

6.5.1 The varieties of causality

My case that EDT and CDT disagree over *EPR* depends crucially on your agreeing with (6.3) that nothing that happens at either receiver affects what happens at the other. Dropping that assumption blocks the derivation of (6.24)–(6.26). For we can no longer rely on (6.15)–(6.18) and their analogues, which asserted the independence of certain conditional chance distributions on the basis of the causal independence that (6.3) describes.

We could be assured of the causal independence of events at the two receivers if what explained the correlations (6.1) and (6.2) was some prior state of the apparatus that was a common cause of both readings. In a well-known analogy due to Bell: if we know that a person likes to wear socks of different colours, then we know in advance that there will be a correlation between what we see when we observe the colour of the left sock and what we see when we observe the colour of the right sock, however causally isolated these observations are. The observations will always reveal socks of different colours, not because either observation causally influences the other but because both are effects of a common cause. In this case it is the person's decision to wear, say, a pink sock on his right foot and a green sock on his left.[9]

So we could assure ourselves that there is no causal influence across the receivers if we were able to explain the correlations (6.1) and (6.2) in similar

---

[8] E.g. Joyce 1999: 149. Some deny that the stipulation is coherent on the grounds that my present action can only be symptomatic of its effects (e.g. Price 2012: 510): for discussion of this see sections 8.1–8.3 below.

[9] The example is from Bell 1981: 139.

fashion. If we are to explain (6.1) in this way then it seems that only one hypothesis will turn the trick. We must suppose that prior to their reaching the receivers some prior state λ of the apparatus determines that the two signals are already in agreement, in the sense that each will produce the same reading for any particular setting of its receiver. For instance, we might write λ = YYN to indicate that *both* signals are such as to give a reading of 'y' at a receiver that is set to 1 or 2 and a reading of 'n' at a receiver that is set to 3. That would explain (6.1) because it would imply that both signals are already set to yield the same reading given the same setting. And, as with the socks, nothing in this explanation would require any causal influence to run between the two measurement events themselves.

What Bell showed, however, was that *unless* there is a correlation between this λ and your choice, on any particular run, of the setting of the receivers, this hypothesis could *not* account for (6.2). More specifically, if λ and your (possibly subsequent) choice of settings of the receivers are uncorrelated, then we should expect that when the receivers are at *different* settings they will give the *same* reading on at least 33 per cent of many runs. This is inconsistent with (6.2).[10]

In the foregoing argument, the events **110**, **101** etc. specifying common conditional chances of readings on settings played the role of values of the parameter λ. What Bell's argument shows in the present connection is there-fore that either (a) the receivers are causally dependent, or (b) the receivers are causally independent but there is a *correlation* between their settings and the conditional chance of getting this or that result given this or that setting. Which of these is true will depend on the true analysis of causation.

Possibility (a) will seem plausible if you take a sufficient condition for causality to be statistical correlation, or regularity conditional on a fixed past state of the world, which on this hypothesis does *not* include a *common* conditional chance distribution λ. For instance, it is true that on any run of the device in which *both* receivers are set to 1, there is a perfect statistical correlation between a 'y' reading on receiver A and a 'y' reading on receiver B. In that case we should have to say that each of these readings causes the other, so that here we have a very tight and rather peculiar causal loop.[11] This would be the correct view on Reichenbach's probabilistic theory of causality in *The Direction of Time*. At least this is so, *if* we prescind from Reichenbach's requirement that the cause be earlier than the effect, for it may be that what happens at either receiver is not in the unequivocal

---

[10] Bell 1964. My argument at sections 6.2–6.3 essentially applies Bell's argument to conditional credences versus credences in conditional chances demanded by crosswise causal independence, in place of observed frequencies versus frequencies demanded by a prior screening-off state.

[11] Skyrms 1984a: 246.

past (the past light cone) of what happens at the other.[12] Similarly, it follows from Suppes's probabilistic account when modified in the same way,[13] from the classic regularity accounts of Hume and Mill and also from what Mackie considers the best refinement of the regularity approach if, again, we ignore the time-biases built into these theories.[14] Possibility (a) will also appear plausible on a counterfactual theory like Lewis's.[15] On (a) the argument for (6.24)–(6.26) collapses. And anyone who is 100 per cent certain of a theory of causation that leads to it will get the same advice from CDT, in any D_i (z) and also in *EPR*, that she gets from EDT.

Many physicists and philosophers have rejected (a) on the grounds that it requires superluminal causality, i.e. influence that travels faster than light, contrary to the special theory of relativity.[16] This is because we could certainly arrange for the two receivers to be so far apart that light could not leave receiver A after its signal has arrived in time to reach receiver B before *its* signal does. It is possible to challenge this argument.[17] But rather than pursuing the point let me just note that there do not seem to be *completely decisive* reasons to accept (a). That fact is enough to support my second argument against CDT at 6.4. As we shall see at section 6.5.2, it also suffices to support the first.

(b) Other analyses of causation support the view that the readings on either receiver are causally independent of the readings and settings on the other receiver. This includes the 'process' theory, according to which causation involves the transfer of some conserved quantity,[18] and also the agency or manipulationist theory, according to which causal connections display how, at least in principle, an agent might intervene upon the cause as a *means* to adjusting the effect.[19] Hausman's theory of causation in terms of non-accidental connections (the n-connection theory) also seems to have this consequence.[20]

---

[12] Reichenbach 1971: 204.   [13] Suppes 1970: 12 ff.

[14] Hume 1975 [1777]: 76. Mill 1952 [1843]: III.v; Mackie 1980: 62. That statement of the 'inus' theory does not explicitly mention time-order. But Mackie holds (ibid.: 82–4) that we should supplement the theory with a requirement that the cause be 'causally prior' to the effect, where A is causally prior to B iff B was not 'fixed' at any time *before* that at which A was fixed (ibid.: 190).

[15] Butterfield 1992b.   [16] See e.g. Jarrett 1984.

[17] Butterfield (1992a: 72–7) challenges it but rejects superluminal causation for other reasons.

[18] Salmon 1984: 171.

[19] Menzies and Price 1993; Woodward 2003. The claim that manipulationism is committed to (b) requires further comment. It is impossible for an agent to manipulate the *setting* on one receiver to produce any desired result at the other receiver: this follows from the 'no-signalling' result that it is impossible to use the device as a superluminal telephone (Bell 1985: 60–1). And it has no clear meaning to say that an agent could by some *other* means manipulate the reading at one receiver to produce a desired reading at the other.

[20] At least this is true of the version of the theory with which Hausman himself proposes to deal with EPR-type cases (Hausman 1999).

According to (b), what explains (6.1) and (6.2) is not that what happens at either receiver influences what happens at the other. Rather there is a correlation between your choice of settings and the pre-existing conditional chances of getting this or that reading conditional on this or that setting. For instance, whenever you set receivers A and B to 1 and 2 respectively, the prior distribution is 010, 011, 100 or 101 three times as often as it is 000, 001, 110 or 111. But your thus manipulating these settings has no *causal* effect on the distribution. *EPR* therefore resembles the Egan–Gibbard cases (section 3.1), in that your actions evince their own pre-existing dispositions to this or that effect.

Philosophers have sometimes rejected this idea out of hand on the basis that it either (i) is incompatible with your freedom[21] or (ii) presumes some incredible conspiracy.[22] Objection (i) itself seems to require two assumptions: (i)(a) that if the setting is not itself causally relevant to the distant reading then your choice of setting and the chance distribution have a common cause; and (i)(b) that a caused choice is not a free choice. But it isn't clear why (i)(a) should be true: why could it not be that here we just have a brute correlation for which *no* causal explanation is available? In any case (i)(b) appears simply to reflect an unargued prejudice against compatibilism.

As for (ii), this objection also appears to depend on (i)(a). Correlations between options, and events that they do not cause, only reveal a conspiracy if there must have been *some* cause in their common past that explains the correlation. But maybe at this level there just *are* correlations between settings and prior events, this being no more mysterious than any causal story that could be invoked to 'explain' it.

Butterfield writes that the immediate antecedents of settings and results (a human finger throwing a switch, a signal entering a receiver) are so disparate that even if we reject a common cause, crosswise setting–result correlations are only credible on the basis of superluminal causation.[23] But if we give up the need for any sort of causal explanation, and hence the temptation to see any conspiracy, why should the mere *disparity* of these (types of) events make a correlation between them incredible? One might just as well argue against the eliminativist about causation (say, Russell) that his position leaves it intolerably mysterious that matches light whenever you strike them.

Maudlin writes:

> *if* a theory predicts a correlation, then that correlation cannot, according to the theory, be accidental. A nomic correlation is indicative of a causal

[21] Bell 1990: 243–4.   [22] Butterfield 1992a: 70–1.   [23] Butterfield 1992a: 76.

connection – immediate or mediate – between the events, and is accounted for either by a direct causal link between them, or by a common cause of both.[24]

But this argument involves a loaded understanding of 'accidental'. If 'accidental according to the theory' just means *not predicted by the theory*, then tautologically no theory predicts accidental correlations. But that hardly entails any causal commitments. If 'accidental according to the theory' means *has no causal explanation according to the theory*, then certainly there are theories that predict 'accidental' correlations. But this, according to their advocates, may reflect the *insight* that we should stop looking for causal explanations at the quantum level.[25] Finally, if we simply define 'causality' in such a way as to be somehow involved in *any* nomic connection, then what we are doing 'is nothing but saying that "connected events are connected" . . . using causal concepts in this case appears then to be a mere labeling devoid of any real physical and philosophical significance'.[26] So there are at least prima facie reasons not to give (b) *zero* credence.

One theory of causation does not appear to entail either option (a) or option (b). This is the 'family resemblance' theory that our concept of causation responds to a variety of pressures.[27] Since, in the present instance, some of these pressures push in one direction and others in the other direction, there may just be no saying whether or not (6.3) is true.

We should distinguish two claims: (i) the claim that there is simply no fact of the matter as to whether the receivers are causally independent; (ii) the claim that there is a fact of the matter but nothing to tell us what it is. Claim (i) is very drastic. If there really *is* such a relation as causal dependence then unless we are prepared to give up on classical logic altogether, it either holds crosswise between located settings and readings, or it doesn't. So either these events are causally dependent or they are causally independent. But classical logic is about the last thing that anyone should give up.[28] And although it can be argued that the family resemblance approach to any concept commits us to eliminativism with regard to it,[29] taking this view about causal dependence means that causalism is done for in any case. According to claim (ii), which is therefore the only plausible alternative to eliminativism, the position is that one of (a) and (b) is true but we do not, and perhaps never can, know which.

---

[24] Maudlin 2002: 90.   [25] Van Fraassen 1991: 372–4.
[26] Laudisa 2001: 229.   [27] Skyrms 1984a: 254.
[28] For discussion of the costs and benefits of this view see Quine 1981.
[29] Forster 2010: 85–7 seems to me to encourage this thought, although that is not the author's explicit intention.

So the causalist who takes the family resemblance view should divide his credence in some way between the hypotheses (a) that the receivers are causally related and (b) that they are causally independent. Since it is of the essence of this view that the pressures that normally push in one direction are in this case ambiguous, anyone who takes it is free to adopt any such division: certainty that the receivers are causally related, certainty that they are causally independent, and any intermediate position, are all equally consistent with the family resemblance theory.

Let me summarize the discussion. What advice CDT gives, in the family of cases $D_i(z)$ and also in *EPR*, turns on whether it is (a) or (b) that tells the truth about this device's causal structure. So I concede that my argument at sections 6.1–6.3 turns on our taking the (b)-theory for granted. What this present objection shows, however, is that we *cannot* take it for granted.

But as long as we cannot be sure that (a) is true, the case still presents difficulties for CDT. In particular, the second argument at section 6.4 still applies. That argument was not that CDT gives the wrong advice, but rather that *what* advice it gives turns on a question of philosophical analysis, in particular on whether we should analyse the causal relation in terms that support (a) or in terms that support (b), or in terms that leave it open. But however that question is ultimately resolved, the fact that CDT makes the practical question how to bet *turn upon it* still constitutes grounds for rejecting CDT.[30]

6.5.2   Mixed beliefs

But the *first* argument, that CDT gives the *wrong* advice in at least one of these cases, is still on the back foot. The reasoning at sections 6.2 and 6.3 only applies to agents who are *certain* that what happens at either receiver is causally irrelevant to what happens at the other. But the foregoing discussion makes it reasonable to give that thesis only *some* credence. It

---

[30] I have in this discussion altogether ignored the 'many-worlds' theory according to which all possible results of any run appear in one of the branches into which the world divides after the readings have been taken (Blaylock 2010: 116–17). One reason for this is that it is hard to see what any decision theory should say about a case in which the uncertainty of each run gets resolved in different ways according to equally real future counterparts of the experimenter: in particular, it is not clear to me why the 'weights' that the many-worlds theory attaches to each branch should play the decision-theoretic role that subjective uncertainties play in cases of 'single branch' decisions under uncertainty (for reasons arising from Price's (2010) comments on Wallace 2007 and Greaves 2007). But in any case and as section 6.5.2 argues, anyone who gives the many-worlds interpretation some credence should still agree that *EPR* drives EDT and CDT apart, as long as he also gives *some* credence to the 'single-world' proposal on which the receivers are causally independent of one another.

Table 6.5 *EPR (α)*

	yy	yn	ny	nn
12hom	α	0	0	α
13hom	α	0	0	α
23hom	α	0	0	α
12het	0	1	1	0
13het	0	1	1	0
23het	0	1	1	0

remains to be shown that CDT gives wrong advice to agents whose beliefs are thus mixed.

Write C for the proposition that the receivers are causally related. So $\neg C$ says that they are causally independent. Consider some decision problem D_i (z) as in Table 6.1. For any such problem, the V-score of betting iihom is $1 - z$, and that of Q is simply zero. And this is true under any hypothesis about the causal structure of the device, since EDT makes recommendations that are independent of the metaphysics of causation and instead depend only on the observed statistics: at least it does if the F-C connection holds. So EDT will recommend iihom in Table 6.1 to any F-C-respecting agent whose credence is divided between C and $\neg C$.

We cannot directly calculate what recommendation CDT makes to such an agent. However it is true even of *such* an agent that CDT *will* recommend quitting in D_i (z) for some $z < 1$ *unless* equation (6.9) holds. Putting together this point with the insensitivity of EDT to (6.9), we can see that the argument against CDT will hold even on the assumption of divided credence, *unless* (6.9) holds. So we may again take forward (6.9).

Next, consider the family of decision problems in Table 6.5. The table defines a decision problem for each $\alpha > 1$. For instance, *EPR (2) = EPR*. What EDT recommends to the 'mixed' agent in any *EPR (α)* depends in the following manner on the precise value of α:

(6.44) $V(12\text{hom}) = \alpha \, Cr \, (yy \lor nn | 12\text{hom}) = 0.25\alpha$

(6.45) $V(12\text{het}) = Cr \, (yn \lor ny | 12\text{hom}) = 0.75$

and similarly for the other 'hom' and 'het' options. So EDT recommends every 'het' option over every 'hom' option if and only if $\alpha < 3$. Again this is quite independent of the agent's $Cr \, (C)$.

What about CDT? Here things are only slightly more complicated. If C is true then the outcomes are causally dependent on the settings.

In particular, we may assume that these causal dependencies reflect the corresponding frequencies, so that for $1 \leq i < j \leq 3$ we have:

(6.46) $Ch\,(yy \vee nn \mid ij) = 0.25$
(6.47) $Ch\,(yn \vee ny \mid ij) = 0.75$

On the other hand, if C is false then the chance distribution of readings on settings will conform to (6.24)–(6.26). So the K-partition now has nine elements: that corresponding to (6.46) and (6.47) above, which we may identify with C itself, and the eight possibilities that (6.24)–(6.26) leave open. Your credences over these other eight hypotheses may be written as follows:

(6.48) $Cr\,(abc) = Cr\,(abc \wedge \neg C) = Cr\,(abc \mid \neg C)\,Cr\,(\neg C)$

– 'abc' here being defined as at (6.31).

Writing c for $Cr\,(C)$, and $Cr_C\,(x)$ and $Cr_{\neg C}\,(x)$ for the marginal distributions $Cr\,(x|C)$ and $Cr\,(x|\neg C)$ respectively, we therefore have the following six expressions for the U-scores of the options for an agent whose credences are divided over whether the receivers are causally isolated:

(6.49) $U\,(12\text{hom}) = 0.25\alpha c + \alpha(1 - c)\,Cr_{\neg C}\,(111 \vee 110 \vee 001 \vee 000)$
(6.50) $U\,(12\text{het}) = 0.75c + (1 - c)\,Cr_{\neg C}\,(101 \vee 100 \vee 011 \vee 010)$
(6.51) $U\,(13\text{hom}) = 0.25\alpha c + \alpha(1 - c)\,Cr_{\neg C}\,(111 \vee 101 \vee 010 \vee 000)$
(6.52) $U\,(13\text{het}) = 0.75c + (1 - c)\,Cr_{\neg C}\,(110 \vee 100 \vee 011 \vee 001)$
(6.53) $U\,(23\text{hom}) = 0.25\alpha c + \alpha(1 - c)\,Cr_{\neg C}\,(111 \vee 011 \vee 100 \vee 000)$
(6.54) $U\,(23\text{het}) = 0.75c + (1 - c)\,Cr_{\neg C}\,(110 \vee 010 \vee 101 \vee 001)$

Now we know by the structurally identical reasoning of (6.32)–(6.42), which applies just as well to $Cr_{\neg C}$ as to Cr, because the former is also a probability distribution, that at least one of the following is true:

(6.55) $2\,Cr_{\neg C}\,(111 \vee 110 \vee 001 \vee 000) \geq Cr_{\neg C}\,(101 \vee 100 \vee 011 \vee 010)$

(6.56) $2\,Cr_{\neg C}\,(111 \vee 101 \vee 010 \vee 000) \geq Cr_{\neg C}\,(110 \vee 100 \vee 011 \vee 001)$

(6.57) $2\,Cr_{\neg C}\,(111 \vee 100 \vee 011 \vee 000) \geq Cr_{\neg C}\,(110 \vee 101 \vee 010 \vee 001)$

For, if all of (6.55)–(6.57) are false then adding together their denials gives a falsehood analogous to (6.42). Suppose without loss of generality that (6.55) is true. If we now write $t =_{\text{def.}} \alpha - 2$, $p =_{\text{def.}} Cr_{\neg C}\,(111 \vee 110 \vee 001 \vee 000)$ and $q =_{\text{def.}} Cr_{\neg C}\,(101 \vee 100 \vee 011 \vee 010)$ then subtracting (6.50) from (6.49) gives:

(6.58) $U(\text{12hom}) - U(\text{12het}) = 0.25c(t-1) + p(2+t)(1-c) - q(1-c).$

(6.55) tells us that $2p - q \geq 0$. It follows that $p \geq 1/3$ since $p + q = 1$. So:

(6.59) $U(\text{12hom}) - U(\text{12het}) > 0$ if $0.25c(t-1) + t(1-c)/3 > 0.$ Hence:

(6.60) $U(\text{12hom}) - U(\text{12het}) > 0$ if $t > 3c/(4-c)$

If $c < 1$ then there is *always* some t strictly between 0 and 1 that satisfies the inequality on the right-hand side of (6.60).

But since $\alpha = t + 2$, this means that if $c < 1$ then we can *always* choose some pay-off α to the hom options, strictly between 2 and 3, on which the causalist will prefer 12hom to 12het (or more generally, some 'hom' option to the corresponding 'het' option). And by (6.44) and (6.45), we know that $\alpha < 3$ guarantees that EDT always prefers any 'het' option to every 'hom' option. So if $c < 1$ – that is, if the agent gives *any credence at all* to the thesis that the receivers are causally independent – then EDT and CDT will disagree over *some EPR (α)* – or at least, over some simplified problem that stands to it as Table 6.4 stands to Table 6.2.

So the objection fails: as long as the agent is not absolutely certain of causal influence between the receivers, it is possible to construct some *EPR (α)* in which EDT and CDT give divergent advice. And any such case will equally support *both* of the arguments against CDT that section 6.4 based upon such 'pure' cases as those in Tables 6.1, 6.2 and 6.4.

CHAPTER 7

The standard Newcomb case

This chapter explicitly shares the involvement with the supernatural that has characterized much of the philosophical literature on CDT since 1970, for it deals directly with the most popular Newcomb case, that involving a being with extraordinary predictive powers as described at section 2.5. Instead of questioning either (a) what sort of mechanisms could possibly underlie these powers or (b) whether any deliberator could rationally believe that they were operative, I shall for most of this chapter just take it for granted that the being could exist; and that if he does, a deliberator would be rational to treat her own choice as evidentially but not causally relevant to the being's prediction of it.

I won't repeat the details of this Newcomb case. But a reminder of the pay-offs (assuming as usual that value equals dollar value) is given at Table 7.1. This chapter considers at greater length the reasons that I mentioned at section 2.5 for taking this case to motivate CDT.

Arguments from the standard Newcomb problem fall under heading (b) in the division of causalist strategies at the start of Chapter 4. That strategy involved (i) countering reasons to dismiss the standard Newcomb problem as irrelevant; (ii) arguing that CDT gets the case right. In connection with (i), section 7.1 discusses whether causalism can rebut or accommodate the deflationary responses to Newcomb's problem. In connection with (ii), section 7.2 argues against the proposal that whether EDT or CDT gets the case right depends on whether we take the predictor to be infallible. Sections 7.3 and 7.4 discuss arguments for O_1 and O_2 respectively.

7.1 Deflationary responses

There are four deflationary reactions to the standard Newcomb case. (i) First: since nobody has experienced anything like it, no intuition about

Table 7.1 *Newcomb's problem*

	S_1: being predicts O_1	S_2: being predicts O_2
O_1: you take one box	M	o
O_2: you take both	M + K	K

the case deserves much weight.[1] So widespread intuition in favour of two-boxing ought not to constitute much evidence against EDT.

But it isn't clear what experience *could* teach anyone about standard Newcomb problems: after all, no party in the dispute is ignorant about what would happen in a real-life Newcomb case. (What would happen is that the one-boxers would get rich.) And besides, it is common enough for an agent to meet a type of decision situation concerning matters beyond her own experience, for instance whether to have children, or whether to try a new kind of pudding. *Here* she may rely on her practical intuitions. So why should it matter that the standard Newcomb case lies outside anyone's experience? Unless the standard Newcomb case is so radically different from anything that anyone has encountered as to make these intuitions worthless – but that needs further argument.

(ii) The second response compares this case to puzzles that seem unanswerable not because they are profound but because they demand semantic stipulation. (a) Consider my stereo: it consists of an amplifier, a streamer, two speakers, a converter and various wires. If I replace one speaker, the converter and some of the wires, is it still the same stereo? (b) I see a chair over there. But when I go to fetch it, it disappears from sight. 'So it wasn't a chair.' But a few seconds later I see it again, am able to touch it and so on. Then it disappears again.[2] Is it really a chair or not?

The situation is not, in either case, that the answer is for some reason too deeply buried to see. The answer is that in both cases you can say what you please, as long as it does not prevent you from seeing the facts.[3] They are as follows: (a) these bits were replaced and those other bits were not and (b) something chair-shaped appears at these times and disappears at those other times. Determining whether the stereo survived, or whether

---

[1] This seems to be one part of Horwich's view. He offers the fact that the Newcomb case is 'farfetched' as grounds for not inferring much from our intuitions about it (Horwich 1987: 178). He also offers the fact that intuitions are divided as separate grounds for not doing so. But as I argue later in this section, it is not clear how much weight the second point can bear.

[2] Wittgenstein 2009 [1951]: section 80.   [3] Wittgenstein 2009 [1951]: section 79.

there really was a chair there, is not a matter of filling any gap in our knowledge of what was happening but rather of laying down the semantic law. Case (a) being so unimportant, and case (b) being so unusual, our normal criteria of stereo-identity and chair-existence simply do not cover them, so we must make up *new* rules at this point, just as one might should an unforeseen situation arise in some game. 'But do we miss them when we use the word "chair"? And are we to say that we do not really attach any meaning to this word, because we are not equipped with rules for every possible application of it?'[4]

This second response makes the same point about the standard Newcomb problem. Here it is our concept of practical rationality that does not cover the case. Rather it fragments into two notions: what CDT prescribes and what EDT prescribes. We might call these notions C-rationality and E-rationality. We *know* that one-boxing is E-rational and that two-boxing is C-rational. But if you want to know which is rational *period* then the answer is that there *is* no answer. Or you can make one up if you like, but don't expect that to illuminate anything about rationality as you understood it all along. Asking which of one- and two-boxing is *really* rational is like asking whether it *really is* the same stereo or whether there *really was* a chair there. The question is not deep but empty.[5]

That style of response is welcome as far as it goes, and it can and should be taken far enough to dispose of many, though certainly not all, 'metaphysical' questions. Certainly if applied to the Newcomb problem it would settle things, at least as far as that problem is concerned, in favour of evidentialism. If the standard Newcomb case is thus empty then it cannot constitute any weighty reason for thinking that causal beliefs are practically indispensable.

But there is a crucial difference between the Newcomb cases as opposed to (a) and (b). In each of those cases it was symptomatic of the emptiness of the question that our intuitions simply leave us in the lurch. One simply doesn't have any strong intuition as to whether or not it was (a) the same stereo or (b) a chair. And that's not surprising: it is because the rules that we internalized leave these questions open that nothing pushes anyone strongly in either direction. But when it comes to Newcomb's problem, the problem is not that anyone's intuitions are silent. As Nozick reports,[6] everyone *has* got a firm and clear intuition about this case. It is just that many of them

---

[4] Wittgenstein 2009 [1951]: section 80.
[5] Quine might have taken this line. See his analogous remarks on personal identity as cited in Parfit 1984: 200.
[6] Nozick 1969: 210.

have an intuition in favour of one-boxing and many of them have an intuition in favour of two-boxing. This is not a sign of an empty question that our pre-existing concept of rationality underdetermines. Rather it is a sign of an interesting and contentful question to which half of us have got the wrong answer.[7]

(iii) A third deflationary response doesn't ignore this clash of intuitions but capitalizes upon it. On this view, the *conflict* of intuitions means that we should be reluctant to conclude very much from any one of them. In particular, it would be rash to infer that CDT gets these cases right, or probably gets these cases right, from the fact that there is a widespread intuition in favour of two-boxing, given that there is also a widespread intuition in favour of one-boxing.[8]

Two points undermine this reply. First: although many people do have strong intuitions in favour of one-boxing, what evidence there is seems to show that there are fewer such people amongst those whose intuitions are more 'tutored'. In particular, whereas the evidence shows a preference for one-boxing amongst people in general, there is a preference for two-boxing amongst philosophers, and a strong preference for two-boxing amongst decision theorists.[9] And the decision-theoretic literature reflects this: even amongst those decision theorists who reject or question CDT, this is often for reasons unrelated to the standard Newcomb problem, in which they endorse two-boxing.[10]

Second: as we'll see at section 7.2, it may be possible to explain at least some of the intuitive pull in favour of one-boxing as arising from psychological mechanisms that cast unfavourable light upon it. If so, then as much of our intuition as survives this argument might seem to speak more unequivocally on behalf of two-boxing. More generally, *untutored* intuition is doubtless subject to a variety of pressures that do not always reflect rationality and which reflection might tend to eliminate. On the optimistic assumption that this is so, it is not the case that the prima facie conflict of intuitions in Newcomb's problem should stop us drawing decision-theoretic lessons from it.[11]

---

[7] Cf. Parfit's (1984: 199–200) comments on teleportation cases etc.

[8] See the figures cited at Chapter 4 n. 48.

[9] In a recent *PhilPapers* online survey, (i) 19.5 per cent of all respondents 'accept or lean towards' one-boxing and 21.9 per cent accept or lean towards two-boxing; (ii) 21.3 per cent of 'target faculty' philosophers accept or lean towards one-boxing and 31.4 per cent accept or lean towards two-boxing; (iii) 21.5 per cent of target faculty listing decision theory as an area of specialization accept or lean towards one-boxing and 61.3 per cent accept or lean towards two-boxing. Accessed from http://philpapers.org/surveys/results.pl on 19 November 2013.

[10] E.g. Jeffrey 1983: 19–20; Egan 2007: 93–6; Arntzenius 2008: 280 ff.

[11] Williamson (2007: Chapter 6) takes this approach towards philosophical intuitions more generally.

(iv) A fourth deflationary response *concedes* that EDT as standardly conceived gets Newcomb cases wrong, but makes light of this on the grounds that nobody will ever face such a case. It hardly counts against a decision theory that it goes wrong on such peculiar and science-fictional hypotheses as those that Newcomb's problem involves.[12] It is hard not to sympathize with this line of objection, which exerts increasing pressure as we pass from the realistic medical, psychological and economic cases in Chapter 4, via speculations about some future deterministic physics (Chapter 5) and impractical but feasible quantum scenarios (Chapter 6), to the present abstractions.

But I do not wish to press the argument, partly because reasonable people differ over how seriously to take the standard Newcomb problem, and partly because although there is at present little evidence that any very reliable predictor of human choice exists or can be constructed,[13] that might change in the future, and it would hardly do to rest decision theory upon some such bet on scientific progress.

Let me finally emphasize that even if we *did* accept any of these four arguments against taking the standard Newcomb problem seriously, it is still possible that realistic versions of the problem exist. For as I argued at section 4.6, there are reasons to take a modified but still realistic version of *Prisoners' Dilemma* to be just such a thing. Something similar might be said about *Betting on the Past* (though not about *Betting on the Laws* or *EPR*, which are not Newcomb problems). For that reason, and for those just outlined in the discussion of (i)–(iv), I shall set to one side these deflationary defences of EDT. Accordingly, the rest of this chapter is primarily addressed to those who take seriously the standard science-fictional presentation of the Newcomb problem. But what I have to say from section 7.3 onwards will also apply to the realistic Newcomb problems, if any do exist.

7.2   The Discontinuous Strategy

Many people find it intuitively attractive to take both boxes in the standard Newcomb problem as I have described it here. After all, the predictor has already put something or nothing in the opaque box, and you can't affect that. Since it will make you 1,000 dollars richer in either case, you might as well take both boxes. This is the so-called 'dominance' argument for O_2, which section 7.4 discusses further.

---

[12] Jeffrey 1983: 25 ('*Prisoners' Dilemma* for space cadets').
[13] In fact even this might not be true, since the Libet experiment may point the way to the construction of such a predictor. See Slezak 2013: 16–17; and section 5.3.2 above.

But its conclusion suddenly becomes less attractive if we suppose the predictor to be *definitely* right. That is: not only does he have a 100 per cent track record of predicting your choices, and anyone else's choices, in the many situations of this type in which he has been involved, but you are for this reason *entirely certain* that he knows what you are about to do. You are as certain of this as you are that $2 + 2 = 4$, or that Monday is the day before Tuesday, or that there is something and not nothing. So you are also certain that you will be a (dollar) millionaire if and only if you take only the opaque box. In short, you are assuming an *infallible* predictor.[14] Given an infallible predictor, many people prefer one-boxing.

This section considers a package that combines these responses. According to it, what you should do depends on your confidence that the predictor is right. If you are *highly* confident but not 100 per cent confident that the predictor is right, you should two-box. But if you are *completely* confident that the predictor is right, then you should one-box. Call this combination the ***Discontinuous Strategy (DS)***. The DS is discontinuous in the obvious sense that the number of boxes it tells you to take is a discontinuous function of your confidence in the predictor. The discontinuity is at the only plausible place in its neighbourhood that it could be.[15] It would clearly be odd and unmotivated suddenly to switch from two-boxing to one-boxing when your confidence in the predictor reaches 90 per cent, or 95 per cent, or 98.21653 per cent. But for some reason it looks more reasonable to switch at 100 per cent. If the DS is right then *both* EDT and CDT get Newcomb's problem wrong. But things are worse for EDT, which in that case gets it wrong 'almost everywhere'.

---

[14] Be careful to distinguish what I am here calling *infallibility*, i.e. that the predictor *is certainly in fact* correct, from *necessary infallibility*: that the predictor is *necessarily* correct. We are given in the story that the prediction was made in the past, and the point of this is supposed to be that that prediction was causally independent of your present choice. It follows that the predictor is at best *contingently* correct; for if he is in fact correct, then had you chosen otherwise he would have been incorrect. (At least this is so on some non-'backtracking' interpretation of that counterfactual: see Lewis 1979a: 33–5; cf. Horgan 1981: 162–5.) So he might easily *have been* wrong even if he is certainly *in fact* correct; that is, infallible in my sense. One unfortunate complication of the literature is that whilst my usage follows the writers that I mainly discuss, some others (e.g. Fischer (1994)) use 'infallibility' for *necessary* correctness. For various explications of 'infallible' see Ledwig (2000: 170 ff.).

[15] The point of 'in its neighbourhood' is to focus attention away from a discontinuity to which Evidential Decision Theory is independently committed. EDT endorses two-boxing if your confidence that the predictor is correct (i.e. your $Cr(S_1 \mid O_1) = Cr(S_2 \mid O_2)$) is no greater than n_0 $=_{\text{def.}} V(M + K) / (V(M + K) + V(M) - V(K))$, here writing $V(X)$ for your news value for the outcome that you get $X, and setting $V(0)$ at 0. But as soon as your confidence rises beyond that level, which is approximately 0.5 assuming linear value for dollars, it abruptly switches in favour of one-boxing. *This* discontinuity is both defensible and irrelevant.

Clearly, whatever makes the DS attractive is not dependent on the precise amounts of money involved. Nor does it depend on the fact that the prizes are monetary rather than of any other sort. Rather, the plausibility of the DS derives from that of the following:

(7.1)   *Certainty Principle (CP)*: In a choice between O_1 and O_2, if you are *certain* that either $O_1 \wedge Z_1$, or $O_2 \wedge Z_2$, where Z_1 and Z_2 are incompatible outcomes, and if you prefer Z_1 to Z_2, then O_1 is rationally superior to O_2.

If we let O_1 and O_2 be as in Table 7.1, and let Z_1 and Z_2 represent the outcomes that you get 1 million dollars and 1,000 dollars respectively, then the CP applies to Newcomb's problem played against an infallible (i.e. certainly in fact correct) predictor. And there it recommends one-boxing. But it does *not* apply to cases where the predictor is even slightly fallible and for this reason it appears compatible with two-boxing there.

So we may represent the DS as arising from the combination of two principles: standard dominance reasoning where the predictor is fallible and the Certainty Principle where he is not. Certainly some writers have endorsed or expressed sympathy for the Discontinuous Strategy on grounds of this sort,[16] and others have at least acknowledged the intuitive pull of the argument.[17] On the other hand, the position seems absurdly capricious.

> [D]oes a proponent of taking what is in both boxes [given fallibility] (e.g., me) really wish to argue that it is the probability, however minute, of the predictor's being mistaken that makes all the difference? Does he really wish to argue that if he knows the prediction will be correct, he will take only the [opaque box], but that if he knows someone using the predictor's theory will be wrong once in every 20 billion cases, he will take what is in both boxes? Could the difference between one in n, and none in n, for arbitrarily large finite n, make this difference? And how exactly does the fact that the

---

[16] E.g. Leeds (1984: 106); Clark (2007: 143–4); Hubin and Ross (1985: 439). Bach (1987: 416) claims that most proponents of two-boxing against fallible predictors back off from this policy against infallible predictors. Leslie (1991: 73–4) says something similar, although he himself is a thoroughgoing one-boxer. More forthrightly, Levi (1975: 173, 175 n. 7) claims that in Newcomb's problem as (under-)described at the start of Chapter 7, we typically do not know what probabilities to apply, and so should follow a strategy like 'maximin', which recommends two-boxing. But if we know that the predictor is in fact perfectly accurate then we should one-box (ibid.: 172–3).

[17] E.g. Nozick (1969: 232; but see immediately below in the main text); Gibbard and Harper (1978: 370); and Seidenfeld (1984: 203). Horgan (1985: 230–1) grounds the present application of the CP in premises about what the agent has or has not the *power* to do. Sobel (1988: 109–11) criticizes this for misusing the word 'power'. But one might take the CP to support the DS quite independently of that semantic question.

predictor is certain to have been correct dissolve the force of the dominance argument?[18]

This raises two questions.

(A) *Could the Discontinuous Strategy be rational?* More generally, could one reasonably apply the CP to cases of infallible predictors and the like whilst accepting dominance reasoning in 'fallible' Newcomb cases?[19]

(B) *If not then why is it attractive?* If the difference 'between one in n, and none in n, for arbitrarily large finite n' does *not* in fact make any difference to what is rational, then why does it *seem* to make all the difference? What accounts for the pull of one-boxing under infallibility, which everybody feels, but nobody seems able to explain?

7.2.1 *Why it is irrational*

Although the CP itself is in my view plausible, what is not plausible is any position that seeks to combine it with the application of causalism to 'fallible' Newcomb problems, as in the DS. To see why, suppose that you accept the DS because you accept the CP, and consider the following variant of the Newcomb problem.

Two-Predictor Problem (2PP): As before, you are facing two boxes, one opaque and one transparent. As before, you can take, and keep what is in, either the opaque box only, or that box *and* the transparent box. As before, the transparent box contains 1,000 dollars. And as before, a predictor determined the contents of the opaque box in the usual manner.

But this time the predictor has been drawn by a weighted lottery from a pool of *two* predictors, the draw itself being stochastically independent of your choice. These predictors, Chas and Dave, are of wildly varying competence. Chas *always* makes a correct prediction. So you are certain that this is what happened if you know he is in charge on this occasion. Dave always gets it *wrong*. Whenever *he* is in charge, it is always and only the *two*-boxers who end up millionaires. So your pay-offs are certainly in fact as in Table 7.2. The lottery that determined the predictor was weighted towards Chas at a rate of $n: 1 - n$ ($0 < n < 1$). So you have credence n that on this occasion Chas was running the show. Should you one-box or two-box?

[18] Nozick 1969: 232; on the interpretation of which see also Sobel 1988: 113 n. 2.

[19] Sobel (1988) gives plausible reasons for thinking that the DS is unmotivated; they are also reasons for a causalist to think that it is irrational. But his reasons are not available to anyone who shares my own, non-causalist perspective. The argument against the DS that follows at section 7.2.1 should be equally appealing to causalists and evidentialists.

Table 7.2 *Two-predictor problem*

	C: Chas predicts	D: Dave predicts
O_1: you take one box	M	0
O_2: you take both	K	M + K

Note that even though you are not certain *who* is doing the predicting, still the Certainty Principle applies to *2PP*. To see this, define $Z_1(n)$ and $Z_2(n)$ as follows:

> $Z_1(n)$: You get a free lottery ticket that pays out 1 million dollars with probability n.

> $Z_2(n)$: You get a free lottery ticket that pays out 1 million dollars with probability $1 - n$ *plus* a 1,000-dollar bonus.

Now $n = Cr(C)$ is fixed and independent of anything that you do. And if you choose O_1 then certainly you will get: 1 million dollars if Chas was picked and nothing if Dave was picked. This is certainly equivalent to the lottery ticket that $Z_1(n)$ describes. Similarly, you are certain that if you choose O_2 then you will get: 1,000 dollars if Chas was picked and \$(1M + 1K) if Dave was picked. *This* is certainly equivalent to the lottery ticket that $Z_2(n)$ describes.

So if you are facing *2PP* then you must already be *certain* that *either* you will choose O_1 and get the prize Z_1, *or* you will choose O_2 and get the prize Z_2. It therefore follows from the CP that you should choose O_1 over O_2 in case you prefer the first prize to the second:

(7.2)   In *2PP* you should choose O_1 and not O_2 if $Z_1(n) \succ Z_2(n)$.

Now clearly for any large enough $n < 1$ it is true that $Z_1(n) \succ Z_2(n)$, i.e. that you, and anyone else who prefers more money to less, would prefer a ticket for a 1 million-dollar lottery at odds of $n: 1 - n$ to a 1,000-dollar bonus plus a ticket for that lottery at the opposite odds, i.e. at odds of $1 - n: n$. After all, would you really exchange (1) this 1 million dollars in gilt-edged securities for (2) 1,000 dollars plus a ticket for a real lottery, one that has a one-in-a-million chance of winning? You would rather keep (1). But (1) is really just a lottery ticket that has a very good chance of winning 1 million dollars, since there is one chance in a million that the issuer will go bust before you can redeem the bonds.

It follows from this fact and (7.2) that the CP recommends one-boxing in the two-predictor case for some n strictly less than 1. Let us choose $n^* < 1$ as our representative such n. Then the CP recommends one-boxing in *2PP* if $Cr(C) = n^*$.

Holding on to the supposition that you follow the DS on the basis of the CP, turn now to the standard Newcomb scenario as in Table 7.1, in which there is *one fallible* predictor. In particular his strike rate, and so also your confidence that he is correct on this occasion, is the quantity n^* from the previous example: that is, $Cr(S_1 \mid O_1) = Cr(S_2 \mid O_2) = n^* < 1$. In this scenario you prefer *ex hypothesi* to two-box, for the DS is the combination of views that prescribes one-boxing against infallible predictors and two-boxing against fallible ones.

But this is inconsistent, for this standard scenario, and the 'variant' involving Chas and Dave, are in fact *one* scenario described in two ways. The one predictor in the standard scenario is a person, *Chas'n'Dave*, who is like all persons an aggregate of momentary person-stages, some of which make predictions. The predicting stages are themselves irregularly distributed across two temporally scattered objects. One of these, Chas, is the aggregate of those of Chas'n'Dave's temporal stages – past, present and future – that make *true* predictions, these forming a proportion $Cr(C) = n^*$ of the stages of Chas'n'Dave that make any prediction at all. The other object, Dave, aggregates those temporal stages that make *false* predictions, these forming the remaining proportion $1 - n^*$ of the predicting stages of Chas'n'Dave.[20]

It was Chas'n'Dave himself, who gets it right n^* times for every $1 - n^*$ times that he gets it wrong, who determined what is in the opaque box facing you now. To say that *is* to say that you are facing a gamble, at odds of $n^*:1 - n^*$, that an infallible predictor rather than a hopeless one determined the contents of that box.

It follows that the CP itself, which was supposed to help motivate the Discontinuous Strategy, actually conflicts with it. For the CP implies that when facing a fallible predictor in Newcomb's problem you should *one*-box as long as your confidence that he has got it right exceeds $n^* < 1$.

---

[20] It is true that in the situation described Chas and Dave are probably not distinct *persons*, although they *may* be, if (a) the Chas-stages are psychologically continuous with one another; and (b) so are the Dave-stages; *but* (c) the Chas-stages are not psychologically continuous with the Dave-stages. Certainly they are not *temporally continuous* persons. But nothing in Newcomb's paradox, or in anyone's intuitions regarding it, or in any significant arguments concerning it, depends on the predictor's counting even as a person, let alone a temporally continuous one. If it could be the market (Broome 1989); an alien being (Nozick 1969: 207); or God (Craig 1987; Resnik 1987: 111), then why couldn't it be a temporally scattered sub-personal aggregate of person-stages?

But this is straightforwardly inconsistent with the discontinuous strategy of one-boxing if and *only* if you are *certain* that the predictor is correct.

This leaves two options. First, we might reject the Certainty Principle altogether. In that case the DS loses all motivation. In particular, nobody who feels any sympathy towards the two-boxing strategy against a fallible predictor has any reason at all to doubt it when the predictor is known or certain to be right. The upshot is a purely causalist approach to fallible *and* infallible versions of the standard Newcomb problem.

Second, we might accept the Certainty Principle, and in particular that it applies, via (7.2), in favour of one-boxing against even a fallible predictor. In fact it supports one-boxing over two-boxing just as long as your confidence in the predictor exceeds some n such that you would pay 1,000 dollars to exchange a $1 - n{:}n$ chance of wining 1 million dollars for an $n{:}1 - n$ chance of winning 1 million dollars. On a natural assumption, your strategy is then indistinguishable from a purely *evidentialist* approach to the standard Newcomb problem.[21] That is my own (as yet unargued) preference. But either way, the Discontinuous Strategy itself has got to go.

7.2.2 *Why it is attractive*

Why then has it seemed so plausible? What difference between infallibility and near-infallibility could account for whatever additional attraction one-boxing exerts upon us, or at least upon persons who are otherwise causalists, when we *know* that the predictor is correct?

I suggest that the DS manifests an illusion of reason in which there is already good reason to believe. This is **certainty effect**: a fixed reduction of the probability of an outcome has more impact when the outcome was initially certain than when it was initially probable.[22]

---

[21] The 'natural assumption' is that your value for a lottery is its expected value. In that case, the CP demands one-boxing just in case $n\,V(M) > (1 - n)\,V(M + K) + n\,V(K)$ i.e. iff $Cr\,(S_1 \mid O_1)\,V(M) > Cr\,(S_1 \mid O_2)\,V(M + K) + Cr\,(S_2 \mid O_2)\,V(K)$ i.e. iff EDT demands it. In particular, if we assume that $V(\text{You get } \$X) = X$ then the CP demands one-boxing iff $nM > (1 - n)(M + K) + nK$ i.e. iff $n > (M + K)\,/\,2M = n_0$ as defined at n. 15 of this chapter.

[22] Although this is not the main message of Hubin and Ross 1985, that paper suggests that the DS turns on a failure to distinguish between (i) supposing that the predictor is correct and then comparing what would happen if you were to one-box with what would happen if you were to two-box, and (ii) comparing what would happen if you were to *one-box against a correct predictor* with what would happen if you were to *two-box against a correct predictor*. The idea is that if we attend to (i), as causalists should, then two-boxing always comes out better by 1,000 dollars. But if we attend to (ii), then one-boxing can seem to do better than two-boxing by 1 million dollars to 1,000 dollars. And the appeal of the DS arises from the fact that causalists mistakenly attend to (ii) when considering the perfect predictor case and so conclude that one-boxing is optimal there. But this appears simply to shift the question: why, we want to ask, doesn't a similar confusion have a similar effect when

The following two problems illustrate the effect. 'Each problem was presented to a different group of respondents. Each group was told that one participant in ten, preselected at random, would actually be playing for money. Chance events were realized, in the respondents' presence, by drawing a single ball from a bag containing a known proportion of balls of the winning colour, and the winners were paid immediately.

Problem 1 [$N = 77$]: Which of the following options do you prefer?
1A. A sure win of $30 [78 per cent]
1B. An 80 per cent chance to win $45 [22 per cent]
Problem 2 [$N = 81$]: Which of the following options do you prefer?
2A. A 25 per cent chance to win $30 [42 per cent]
2B. A 20 per cent chance to win $45 [58 per cent].'

The proportion of each group choosing each option is indicated in square brackets.[23]

Looking at these data, what springs out is that whereas 1A is much more popular than 1B, 2A is slightly *less* popular than 2B. This suggests that many people would prefer 1A to 1B *and* 2B to 2A. But this is inconsistent. To see this, consider:

Problem 3: Tomorrow one out of you and three others will be chosen at random to face Problem 1. But you must decide *today* what option to take if you happen to get picked. Which of the following options do you prefer?
3A. If you are picked to face Problem 1 take option 1A.
3B. If you are picked to face Problem 1 take option 1B.

Anyone who chooses 1A in Problem 1 should prefer 3A to 3B in Problem 3. After all, if 1A is the right option today then it is the right option tomorrow – the passage of time cannot make it any less right. And if 1A is the right option in Problem 1 then it is the right option *in Problem 1* whether or not you do in fact face it. That it is 3:1 that you *won't* face it does nothing to change the rationality of taking 1A *if* you face it.

But anyone who chooses 2B in Problem 2 should prefer 3B to 3A in Problem 3. After all, Problem 3 *is* a version of Problem 2, for choosing option 3A in Problem 3 gives you a 25 per cent chance of winning thirty

---

the predictor is imperfect but still very good? And in any case, it isn't necessary to formulate the causalist argument in these counterfactual terms. All that is necessary (if anything is sufficient) to make two-boxing seem compelling is that (a) the content of the prediction is causally independent of your choice, and (b) two-boxing does better than one-boxing on either hypothesis about that content. *This* reasoning is not open to any such counterfactual fallacy as the confusion of (i) and (ii). But it too suffers a peculiar loss of force on the hypothesis of infallibility.

[23] Kahneman and Tversky 1981: 30 with trivial alterations.

dollars and choosing option 3B gives you a 20 per cent chance of winning forty-five dollars. (We can assume that in Problem 3, as in Problems 1 and 2, real money is involved one time in ten.)

So anyone who chooses option 1A in Problem 1 *and* option 2B in Problem 2 is being practically irrational in a way that would show up on confrontation with Problem 3. But the data suggest that probably there are many such people.

The explanation for this lapse of rationality is, as I said, the 'certainty effect', a phenomenon that Allais pointed out in his 1953 paper and which has since been well documented in humans and animals.[24] To repeat: it is that a fixed reduction of the probability of an outcome has more impact when the outcome was initially certain than when it was initially probable.

It is easy to see this effect at work in the experimental data related to Problem 1 and Problem 2. As we move from Problem 1 to Problem 2, the probability of getting thirty dollars under the 'A' option in each problem falls by the same factor as the probability of getting forty-five dollars under the corresponding 'B' option (i.e. both fall by three-quarters). But this has a greater negative impact on the attractiveness of the 'A' option than on that of the 'B' option, since 1A is *more* popular than 1B, but 2A is *less* popular than 2B.

Now apply this to the DS in Newcomb's problem. What makes the problem so troublesome is that two intuitive motivations that normally work together are here in conflict. These motivations are (i) facts about conditional probabilities, (ii) facts about causal dependencies. That is: we may rank options by their V-scores, which depend on (i); or by their U-scores, which depend on (ii). Of course in most everyday situations, but not in the Newcomb case, these rankings coincide.

I suggest that what drives each person's intuition about Newcomb cases of either type (those involving a fallible predictor versus those involving an infallible predictor) is some weighted combination of these motivations. Persons in whom (i) *always* predominates are intuitive *evidentialists* about decision theory. Persons in whom (ii) *always* predominates are intuitive *causalists* about decision theory. For this reason I'll call (i) the E-motive and (ii) the C-motive. I suspect that most people's intuitive response to the Newcomb problem is the net resultant of some weighted combination of the E-motive and the C-motive.[25]

[24] E.g. Shafir et al. 2008.
[25] So too do Joyce (1999: 154); and Nozick (1993: 41–9). As evidence of the fact that most of us are thus intuitively mixed, consider the fact that the intuitive pull of one-boxing increases with the ratio of the potential amount in the opaque box to the known amount in the transparent box. I

Now notice that the certainty effect predicts abrupt variation in the strength of the E-motive as we move from the case where the predictor is *infallible* to the case where he is just *very* accurate. In particular consider these three cases, where in (b) and (c) we take N to be the same very large quantity:

(a) Newcomb's problem with an infallible predictor
(b) Newcomb's problem but the predictor has a strike rate of $(N-1)/N$
(c) Newcomb's problem but the predictor has a strike rate of $(N-2)/N$

The C-motive to two-box remains equally strong in all cases. In (a) the E-motive to one-box is maximally strong: here the conditional probability of being a millionaire given that you one-box is 1. In (b) that conditional probability has fallen to $(N-1)/N$ and in (c) it has fallen further, to $(N-2)/N$.

The certainty effect predicts that the move from (a) to (b) has a greater (negative) impact on the strength of the E-motive than the move from (b) to (c).[26] Equivalently, the move from (b) to (a) has a greater *positive* impact on the E-motive than what *should* rationally be the equally significant move from (c) to (b). So if the certainty effect is strong enough then you would expect that at least in some subjects, the E-motive might jump discontinuously in strength between (b) and (a), and so be strong enough to outweigh the C-motive in (a) but not in (b). And that, finally, might account for the intuitive appeal of the Discontinuous Strategy – which, as I have already argued, is in fact inconsistent with the only principle (namely the CP) that even looks like supporting it.[27]

---

can report at least anecdotally that more people are inclined towards one-boxing when the former is supposed to be, say, $1 billion, and the latter is supposed to be one cent.

[26] It might concern you that I am not comparing like with like: in the move from option 1A to option 2A the associated chance fell by the same multiplicative *factor* as in the move from option 1B to option 2B. But in the move from case (a) to case (b) the associated probability fell by the same additive *increment* as in the move from (b) to (c). But if N is large then this difference is insignificant, since in that case $((N-1)/N)^2 \approx (N-2)/N$, and so the *factor* by which the predictor's strike rate (b) exceeds that in (c) is roughly equal to the factor by which the strike rate in (a) ($=1$) exceeds that in (b).

[27] I should here explicitly acknowledge what will anyway be obvious, that the psychological source of this intuitive illusion is really an empirical matter that could only be decisively settled by experiment. But this section was written entirely from an armchair. Let me give two excuses for that. First: to my knowledge nobody has ever suggested anything like this explanation for the attractiveness of the DS. So even from the armchair it is possible to advance matters by putting it forward. Second: there is in the nature of infallible Newcomb cases some difficulty in testing the relative strength of E-motives and C-motives. It may be hard to get anyone to believe that somebody else is a very accurate predictor of her choice. How could you ever convince someone that the predictor is *infallible*? Perhaps this explains the dearth of even aspirationally rigorous studies of intuition in this

There are therefore two points that I wish to take forward. The first is that the Discontinuous Strategy, which recommends one-boxing if and only if the predictor is infallible, is irrational and should be rejected on all hands. What this means is that we need only consider the 'pure' evidentialist and causalist views, i.e. respectively one-box whether the predictor is infallible or just very reliable, or two-box in both types of case. It follows in turn that I needn't consider separately the case of fallibility and that of infallibility. It will do to consider just one of these two versions of the standard Newcomb problem.

But the second point is that DS may be attractive because of the certainty effect, i.e. because we irrationally overweight certainty. If so, our intuitions regarding infallible predictors are less trustworthy than those directed at 'fallible' Newcomb cases. For this reason, I will from now on focus on those slightly less tidy versions of the Newcomb problem in which the predictor is known to be highly reliable but in which he is nowhere assumed to be perfect.

7.3   The case for one-boxing

I turn now to arguments for one-boxing and two-boxing. I cannot hope to discuss all of the arguments that have appeared since Nozick introduced the problem to philosophers. Instead I shall focus on those on either side that have seemed most plausible. In fact I'll only discuss at length *one* argument for one-boxing, this being the one that has, for good reason, been most prominent in the literature. But first let me mention two others that are at least prima facie plausible.

The first is a sort of 'slippery slope' argument. Starting from the premise that one-boxing is rationally optimal in the 'infallible' case, i.e. where the predictor is *certainly* right, it then asks why it should make any difference if we make the predictor only very slightly fallible; and if it doesn't, then why isn't one-boxing also optimal when the predictor is only *highly* reliable?[28] I have already stated my reason for setting this argument aside, namely the suspicion that what drives our intuitions about the 'infallible' case is a bias that generates demonstrably irrational judgements elsewhere.

The second line is the 'meta-argument' for one-boxing. If any good argument for two-boxing exists then the predictor could foresee its use by the deliberator. So such an argument could never pay out more than

special case (Anand 1990 being the only one known to me). In any case it makes it plausible that the present enquiry is best pursued, because only pursuable, from the armchair.

[28] Horgan 1981: 169; Seidenfeld 1984: 203–4.

1,000 dollars. Whereas *any* argument for one-boxing, if foreseen when used, would pay out 1 million dollars every time it was used.[29] Whether or not this argument vindicates one-boxing turns in part on what is supposed to count as a 'good argument', this itself being something on which evidentialists and causalists may disagree.[30] In particular, whether its 'paying out 1 million dollars' speaks decisively in favour of *either* an argument *or* an actual strategy is an issue that arises more sharply in connection with the argument for one-boxing that I now discuss at greater length.

7.3.1 Why Ain'cha Rich?

> The superiority of Bayesian methods is now a thoroughly
> demonstrated fact in a hundred different areas. One can argue
> with philosophy; it is not so easy to argue with a computer
> printout, which says to us: 'Independently of all your philosophy,
> here are the facts of actual performance'.
>
> E. T. Jaynes[31]

The disagreement between EDT and CDT runs so deep that very little counts as an uncontentious basis on which to adjudicate between them. One principle that *looks* uncontentious is that if, of two options O_1 and O_2, O_1 has a foreseeably better expected return than O_2, then instrumental rationality should prefer O_1 to O_2.

That principle clearly favours one-boxing in Newcomb's problem. If your confidence in the predictor's correctness is n, where $n \gg 0.5$ (but in accordance with section 7.2.2 we hold $n < 1$), the expected return (ER) to your options is as follows:

(7.3) $\text{ER}(O_1) = nM$

(7.4) $\text{ER}(O_2) = (1 - n)(M + K) + nK = K + (1 - n)M$

So one-boxing has a better expected return than two-boxing if $n > 0.5 + K/2M$, which is guaranteed by the assumption that $n \gg 0.5$.

If your confidence in the predictor's abilities reflects the actual statistics of his performance, repeated observation of the Newcomb scenario will make this argument apply. One-boxers will very often end up with 1 million dollars whereas two-boxers will just as often leave with 1,000 dollars. Anyone who repeats the one-boxing policy will make millions upon millions. His fortune will soon outstrip that of the persistent two-boxer by a factor of 1000 to 1 ($= M / K$). We can even imagine the wealthy

---

[29] Bach 1987.   [30] Sobel 1989.   [31] Jaynes 2003: xxii.

one-boxer explaining patiently to the poor causalist just what he needs to do to become a millionaire – which is after all very simple: he just needs to one-box. And we can imagine the causalist repeatedly doing just the opposite, with unfortunate results that were foreseeable in advance to *both* of them.

That story is not so much a separate argument as a dramatization of the fact that everyone on both sides can foresee that EDT does better than CDT. It lends vividness to Arntzenius's comment on that fact:

> In a Newcomb type case evidential decision theorists will, on average, end up richer than causal decision theorists. Moreover, it is not as if this is a surprise: evidential and causal decision theorists can foresee that this will happen. Given also that it is axiomatic that money, or utility, is what is strived for in these cases, it seems hard to maintain that causal decision theorists are rational.[32]

One premise of this argument as stated is that evidential decision theorists will be richer on average than causal decision theorists. That is perhaps not the best way to put it. Disputes between CDT and EDT are not about the relative welfare of *theorists* who advocate those theories. They are about the relative return to the *options* that those theories recommend, whether the actor in question is himself a self-conscious causalist, a self-conscious evidentialist, or, like the majority of people to whom decision-theoretic recommendations should also apply, someone who has never heard of either.

So the key premise is better put like this. The option that EDT recommends in a Newcomb-type situation has a better expected return than the option that CDT recommends there. Making this amendment and affixing Lewis's title for it,[33] we have the following argument:

Why Ain'cha Rich? (WAR)

(7.5) The expected return to one-boxing exceeds that to two-boxing (by (7.3), (7.4)).

(7.6) Everyone agrees on (7.5) (premise).

(7.7) Therefore one-boxing foreseeably does better than two-boxing (by (7.5), (7.6)).[34]

---

[32] Arntzenius 2008: 289. [33] Lewis 1981b. See also Gibbard and Harper 1978: 371.

[34] Of course there is *a* sense in which, compatibly with (7.5) and (7.6), one-boxing does *not* foreseeably do better than two-boxing. One-boxing does foreseeably worse than two-boxing in the sense that on *any* particular encounter with a Newcomb problem, a one-boxer *would* have done better to have taken both boxes. In this 'counterfactual' sense of 'foreseeably better', two-boxing is foreseeably better. So distinguish that *counter*factual sense of 'foreseeably better' from the sense in which it

(7.8) Therefore CDT is committed to the foreseeably worse option for anyone facing Newcomb's problem (by (7.7)).

(7.9) Therefore CDT gives the wrong recommendation in the standard Newcomb problem (by (7.8)).

It is clear enough that the premises are true. What is at issue between proponents of this argument and defenders of CDT is the validity of the argument, and in particular whether the conclusion (7.9) really follows from (7.8). That is, whether an option that has a foreseeably *worse* expected return than another that is also available should for that reason be discarded. That principle, which I just called apparently uncontentious, is in fact contentious, because it *has* been reasonably contested, on bases that I'll now consider.

7.3.2 *Joyce on* WAR

Joyce frames the causalist response as a reply to the title question. His evidentialist, Irene, has just won a million. His causalist, Rachel, has ended up with just 1,000 dollars. Irene now asks Rachel, 'If you're so smart why ain'cha rich?' Rachel replies:

> I am not rich . . . because I am not the kind of person the [predictor] thinks will refuse the extra $1,000. I'm just not like you, Irene. Given that I know that I am the type who takes the transparent box, and given that the [predictor] knows that I am this type, it was reasonable of me to think that the $1,000,000 was not in the opaque box. The $1,000 was the most I was going to get no matter what I did. So the only reasonable thing for me to do was to take it.[35]

Implicit in this line of reply are three assumptions. (a) There are two 'types' of person: the 'C-type' that takes both boxes, and the 'E-type' that takes only one. (b) The predictor bases the contents of the opaque box on his knowledge of the agent's type. (c) In advance of choosing, the agent *knows* her type (for note, Rachel is presented as knowing before choosing that the 1 million dollars is not in the opaque box). These assumptions are not present in standard formulations of the Newcomb problem. Worse, they are at odds with its basic point.

---

means: foreseeably has a better *actual* expected return. In that second sense, which is what I intend, all parties will agree that one-boxing does foreseeably better than two-boxing given that the predictor is foreseeably accurate. What is at issue is not *that* point, but whether anything follows *from* that point about the superiority of EDT.

[35] Joyce 1999: 153 with trivial changes, e.g. Joyce's 'predictor' is supposed to be a psychologist.

To see this, note that if (a)–(c) are true then both CDT *and* EDT advise Rachel to take both boxes. If Rachel already knows that she is a C-type then what she now does will be evidentially relevant *neither* to her being one, *nor* therefore to her confidence that the predictor has identified her as one. So against this background of knowledge, and assuming a confidence of n that the predictor has correctly identified her type, her credence function must satisfy:

(7.10) $Cr(S_1 \mid O_1) = Cr(S_1 \mid O_2) = Cr(S_1) = 1 - n$

(7.11) $Cr(S_2 \mid O_1) = Cr(S_2 \mid O_2) = Cr(S_2) = n$

So $V(O_1) = (1 - n)M < V(O_2) = K + (1 - n)M$. So on these assumptions, even the *evidential* theory recommends two-boxing. A similar argument covers Irene: knowing that she is an E-type, she has a fixed confidence n that the predictor knows this and so has put 1 million dollars in the opaque box. But this gives her every reason, even by evidentialist standards, to take both boxes. On assumptions (a)–(c) therefore, EDT and CDT make the same prescription and the decision situation no longer sustains the contrast that is supposed to motivate CDT.

If we drop all three assumptions everything looks simpler. There need be no 'type' of person that always, or nearly always, sticks to one or the other strategy. It may be that both Rachel and Irene have been in this situation in the past on many occasions and that both have taken the transparent box on about 50 per cent of them. Nevertheless, each of Rachel and Irene is very confident, in advance of deciding what to do, that whatever she *does in fact* do on *this* occasion is what the predictor has predicted that she would do on this occasion. As it happens then, Irene takes only the opaque box in which she finds 1 million dollars. Rachel takes both boxes, finds nothing in the opaque box and keeps the 1,000 dollars in the transparent box. Irene to Rachel: If you're so smart then why aren't you rich?

The last part of Rachel's reply still gestures at a plausible point independently of (a)–(c). I mean her remark that 'The 1,000 dollars was the most I was going to get no matter what I did. So the only reasonable thing for me to do was to take it.' Consider Rachel's situation *before* her choice but *after* the predictor's. The predictor has already put nothing in the opaque box (and of course 1,000 dollars in the transparent box). However carefully and intelligently Rachel had then considered the situation, she was not in any position to have affected that. So *in* that situation she did as well by taking both boxes, as she possibly could have done. Her situation at the time of her choice was one in which she could not have made more than 1,000 dollars. Nor could anyone else in that situation, however 'smart'.

So notwithstanding her false assumptions about the nature of the problem, Joyce's Rachel does have a reply to the question 'Why ain'cha rich?' She is as rich as she could have been given the situation that she was in. If that reply really does rebut the argument for which the question stands, then it must show that the step from (7.8) to (7.9) is invalid, for nothing in this reply shows any inadequacy in the preceding steps or premises. The objection to the argument is, I suppose, that even a foreseeably-worse-than-some-alternative option could be rational. For it is *compatible* with its being thus foreseeably worse that it still is best for the agent *given the situation that she is presently in.*

But it looks beside the point. *Why* exactly does it speak for two-boxing that it gets you 1,000 dollars more than one-boxing *would have done, if you had* one-boxed in the same situation as that in which you actually two-boxed?

One thing that does follow from it, or rather from Rachel's knowledge of it, is that she won't *regret* what she did. Regret is a modal notion in the sense that to regret acting in some way is (at least) to think that you *would* have done better, or got more of whatever it is that you value, if you *had* acted in some other way. And Rachel knows that she would not have done any better than she actually did if she had one-boxed. It is equally true and predictable *ex ante* that Irene *will* regret what she did if she one-boxes. For whilst she is happy enough to have ended up with the 1 million dollars associated with one-boxing, she also knows that she'd have ended up with 1,000 dollars more than that if she had two-boxed instead.

Irene can and should grant all of that. But she can and should insist that it simply isn't relevant. The agent who is now deciding whether to take one box or two knows, or at least has arbitrarily high confidence short of certainty, that at the *actual* world one of two things is in fact going to happen. *Either* she is going to take only the opaque box and end up with 1 million dollars, *or* she is going to take the transparent box as well and end up with just 1,000 dollars: $(O_1 \wedge S_1) \vee (O_2 \wedge S_2)$. And it is up to her which one does happen, in the sense that for each disjunct of this disjunction, the agent has it in her power to realize a conjunct of that disjunct. All of this is given in advance. And since she is only concerned about money, it is clear enough what she should do.

Rachel is pointing out that something else is also given in advance: if the agent takes both boxes then she will get 1,000 dollars more than she *would have* got had she taken only the opaque box, and if she takes only the transparent box then she will get 1,000 dollars less than she would have got if she had taken both boxes. Equivalently, she is pointing out that at the actual world any two-boxing agent will certainly get 1,000 dollars more than her merely possible one-boxing *counterpart*, and that at the actual world any *one*-boxing agent will certainly get 1,000 dollars *less* than her *two*-boxing counterpart.

But neither Irene nor any other agent cares, when choosing, about whether she does *better* than anyone, actual or possible, counterpart or not. Nor does she care about whether she will consequently regret her choice. She only cares about her terminal wealth. So Rachel's point is irrelevant. Given the set-up of the problem, it simply makes no difference that one-boxing has a foreseeably regret-laden outcome in which she does worse than her counterpart, and two-boxing a foreseeably regret-free outcome in which she does better than her counterpart. What matters is only that she is *richer* in the first outcome than in the second.

The reply that I have made here on Irene's behalf does, it is true, depend on a certain way of understanding Rachel's point. I am implicitly equating the point, that two-boxers are making as much money as they could in their situation, with the comparative judgement that any two-boxer does better than her counterpart. And I am saying that that point is irrelevant because the agent's aim was not to do better than somebody else, even her own counterfactual self. It was the completely non-invidious aim of making money for her *actual* self. So perhaps Rachel or Joyce would object that that is a misunderstanding: 'making as much money as *you* could have made in your situation' is not to be understood as 'making more money than your counterpart', but is rather a primitive modal judgement.[36]

This counter-reply itself raises a further question: if the modal judgement is taken as primitive, and so not reducible to any other sort of judgement, then an account is both owing and impossible to give of what it is about such a judgement that gives it any rational traction upon our *actual* choices. And this point in the discussion probably represents a fundamental clash of intuitions. Causalists *just intuit* that judgements about how things *would* be if you *were* to act otherwise are relevant to practical rationality, even to the point (as here) where they outweigh overwhelming

---

[36] One motivation for this objection is in the style of Kripke's 'Humphrey' objection to counterpart theory: Humphrey *legitimately* cares that *he* might have won the 1968 US presidential election, even though he *doesn't* care, legitimately or otherwise, about whether somebody *like* him – that is, some counterpart of him – won or might have won (Kripke 1980: 45 n. 13). For objections to that objection see McFetridge 1990: 143–4.

Table 7.3 *Yankees v Red Sox*

	R: Red Sox win	Y: Yankees win
BR: bet on Red Sox	2	− 1
BY: bet on Yankees	− 2	1

empirical evidence (Jaynes's 'computer printout') that paying attention to these things consistently does worse than ignoring them.[37]

At section 7.4.3 I'll offer independent reasons to think that this intuition is in any case *inconsistent* with CDT. But in connection with *WAR* itself, my conclusion at this point is that Joyce does not refute the argument. His discussion only points to a reason for thinking it invalid, for which there seems no evidence at all.

However, there is certainly more to say as regards *WAR*, and in particular its terminal step. Notwithstanding his apparent endorsement of its spirit as cited above, a well-known argument of Arntzenius seems to make this step unavailable. In particular, he argues that even the *evidentialist* must accept that having a foreseeably worse expected return than some alternative is no grounds at all for rejecting an option.

7.3.3 *Arntzenius on* WAR

Arntzenius claims that if *WAR* works against CDT then an exactly parallel argument works against EDT. So the evidentialist is in no position to defend one-boxing on the basis of *WAR*. Here I describe and then criticize that parallel argument.

> **Yankees v Red Sox:** The Yankees and the Red Sox are going to play a lengthy sequence of games. The Yankees win 90 per cent of such encounters. Before each game Mary has the opportunity to bet on either side. Table 7.3 summarizes her pay-offs on every such occasion as well as our abbreviations for the relevant options and events. Just before each bet, a perfect predictor tells her whether her next bet is going to be a winning bet or a losing bet. Mary knows all this.

---

[37] Thus e.g. Joyce's (1999: 154) discussion of this argument concludes that '[t]he one thing we causal decision theorists are committed to is that considerations of efficacy [which Joyce spells out in modal terms at ibid.: 161 ff.] should always win out in this tug-of-war [against evidence about what actually works] when the issue is one of deciding how to act'. And Lewis (1981c) appears to concede that the weight that causalists attribute to counterfactual considerations ('If I took only one box, I would be poorer by 1,000 dollars than I would be after taking both') is a basic feature of that position.

What does EDT recommend? Suppose the predictor says: 'Mary, you will win your next bet.' Then the news value $V_W(BR)$ of betting on the Red Sox is:

(7.12) $V_W(BR) = 2Cr(R \mid BR \wedge \text{Win}) - Cr(Y \mid BR \wedge \text{Win}) = 2$

And the news value $V_W(BY)$ of betting on the Yankees is:

(7.13) $V_W(BY) = -2Cr(R \mid BY \wedge \text{Win}) + Cr(Y \mid BY \wedge \text{Win}) = 1$

So if Mary knows she will win her next bet then her EDT-rational bet is on the Red Sox.

Suppose the predictor says: 'Mary, you will lose your next bet.' Then the news value $V_L(BR)$ of betting on the Red Sox is:

(7.14) $V_L(BR) = 2Cr(R \mid BR \wedge \text{Lose}) - Cr(Y \mid BR \wedge \text{Lose}) = -1$

And the news value $V_L(BY)$ of betting on the Yankees is:

(7.15) $V_L(BY) = -2Cr(R \mid BY \wedge \text{Lose}) + Cr(Y \mid BY \wedge \text{Lose}) = -2$

So if Mary knows she will *lose* her next bet then her EDT-rational bet is also on the Red Sox. So Mary's EDT-rational bet is going to be on the Red Sox *for every game*.

> So Mary will always bet on the Red Sox. And, if the Yankees indeed win 90 per cent of the time, she will lose money, big time. Now, of course, she would have done much better had she just ignored the announcements, and bet on the Yankees each time. But, being an evidential decision theorist she cannot do this.[38]

It is easy to see that she would do better to bet on the Yankees. The expected returns to betting on the Red Sox and the Yankees are respectively:

(7.16)   ER $(BR) = 90\%. -1 + 10\%. 2 = -0.7$
(7.17)   ER $(BY) = 90\%. 1 + 10\%. -2 = 0.7$

It is also easy to see that CDT recommends betting on the Yankees every time. Win or lose, Mary's bet on any game is causally irrelevant to its outcome. So by (2.34), the causalist's evaluations of those bets are as follows:

(7.18) $U(BR) = V(R \wedge BR) Cr(R) + V(Y \wedge BR) Cr(Y) = -0.7$
(7.19) $U(BY) = V(R \wedge BY) Cr(R) + V(Y \wedge BY) Cr(Y) = 0.7$

So the causalist bets on the Yankees every time; and he makes an expected 70 cents per game. 'So', Arntzenius concludes, 'there are cases in which causal decision theorists, predictably, will do better than evidential decision theorists'.[39]

This is a parity argument against *WAR*, or at least against its use to defend EDT. If *WAR* works against CDT, then this parallel argument works against EDT. In line with the amendment that I proposed at section 7.3.1 to Arntzenius's formulation of *WAR*, I suggest that we make the argument concern options rather than persons. So put, it runs as follows:

Yankees

(7.20) The expected return to betting on the Yankees exceeds the expected return to betting on the Red Sox (by (7.16), (7.17)).

(7.21) Everyone agrees on (7.20) (premise).

(7.22) Therefore it is now foreseeable that betting on the Yankees will do better than betting on the Red Sox (by (7.20), (7.21)).

(7.23) Therefore EDT is committed to what is now the foreseeably worse option for Mary (by (7.12)–(7.15), (7.22)).

The dialectical position is now as follows. The evidentialist might think that *WAR* is an argument for one-boxing in Newcomb's problem. But it looks as though *she* is in no position to say that. A precisely parallel argument, namely *Yankees*, gives just the *same* reason for thinking that EDT gets it wrong in *Yankees v Red Sox*. In short, *WAR*-type arguments cut both ways if they cut either way. So *they* cannot motivate a preference for EDT over CDT.

My objection to this turns on the fact that *Yankees* involves a shift in epistemic perspective that Arntzenius's argument ignores. *Yankees* therefore suffers from flaws that do not affect *WAR*. So one can consistently maintain the latter against CDT whilst denying that the former has any weight against EDT. We can show this by examining arguments in which the relevant flaw in *Yankees* appears more clearly.

> *Check-up*: Every Monday morning everyone has the chance to pay one dollar for a medical check-up at which the doctor issues a prescription if necessary. Weeks in which people take this opportunity are much more likely than other weeks to be weeks in which they fall ill. In fact on average, 90 per cent of Mondays on which someone *does* have a check-up fall in weeks when he or she is ill. And only 10 per cent of Mondays on which someone *doesn't* go for a check-up fall in weeks when he or she is ill. There is nothing surprising or sinister about this. It is just that one is more likely to go for a check-up when one already has reason to think that one will fall ill.

[38] Arntzenius 2008: 289–90.   [39] Arntzenius 2008: 290.

Table 7.4 *Check-up*

	W: well this week	¬W: ill this week
C: check-up	1	0
¬C: no check-up	2	−1

All weekend you have suffered from fainting and dizzy spells. You're pretty sure that something's wrong. Should you go for the check-up on Monday morning? Clearly if you *are* ill this week, it will be better to have the prescription than not, so the check-up will have been worth your while. But if you are *not* ill this week then the check-up will have been a waste of time and money. Your pay-offs are therefore as stated in Table 7.4, which also gives my abbreviations for the relevant events and options. Given this table and the statistical facts, we may calculate the expected return to going and to not going for a check-up:

(7.24)   ER $(C) = 10\%. \ 1 + 90\%. \ 0 = 0.1$
(7.25)   ER $(¬C) = 90\%. \ 2 + 10\%. \ –1 = 1.7$

So the expected return to not going for a check-up exceeds that of going for a check-up. We may therefore construct the following argument against going for a check-up:

Why Ain'cha Well?

(7.26)   The expected return to *not* going for a check-up exceeds the expected return to going for a check-up (premise: from (7.24), (7.25)).
(7.27)   Everyone agrees that (7.26) is true (premise).
(7.28)   Therefore going for a check-up is now a foreseeably worse option for you than not going for one (by (7.26), (7.27)).

Should you then not go for your check-up? That would be insane: *of course* you should go, given the dizzy spells etc. So where has the argument gone wrong?

It has gone wrong in the inference from (7.26) and (7.27) to (7.28). Taken over *every* opportunity for a check-up for *anyone*, it is true that those opportunities that are taken shortly precede illness much more often than those that are not taken. But *you* should not foresee the outcomes of your *present* options on *this* basis. What you should rather use are the expected returns to your options *given what you now know about yourself.* That is: you should compute the expected returns to C and $\neg C$, not amongst all opportunities for check-ups but amongst *occasions on which the subject is suffering from your symptoms.* You should look at what happens to people when they are suffering from fainting and dizziness. Is subsequent illness *amongst these people on these occasions* any more frequent amongst those who go for check-ups than amongst those who do not? Common sense suggests that for *them,* the subsequent incidence of illness is high in both groups and that it is equal in both groups. In that case it is easily verified that:

(7.29)   Amongst people with the symptoms that you now have, the expected return to going for a check-up exceeds that of *not* going for a check-up.

So *for you, now,* going for a check-up is foreseeably the *better* option.

The fallacy of *Why Ain'cha Well* is that of applying an overly broad statistical generalization to a single case. The generalization is overly broad because it is not limited to cases that resemble yours in relevant respects that you know about. Knowing that you are suffering from dizziness and fainting, the statistical generalization that you should apply to yourself is not (7.26). The appropriate generalization covers only that sub-population that resembles your present stage in that respect, i.e. (7.29). Applying (7.26) rather than (7.29) to yourself involves a failure to consider evidence that is both relevant and available.

Why Ain'cha Rich does not make *this* error. To infer (7.7) from (7.5) and (7.6) is not to apply an *overly* broad statistical generalization. Anyone facing Newcomb's problem has *no* evidence that relevantly distinguishes him now from anyone else whom the statistical generalization (7.5) covers; that is, all other persons who ever face this problem.[40] So applying (7.5) to anyone facing Newcomb's problem is not illegitimate in the way that applying (7.26) to your present stage *is* illegitimate.

---

[40] In the medical and other everyday 'Newcomb cases' it is for the most part false that the agent has no evidence that relevantly distinguishes him from anyone else facing the problem, so in these cases *WAR* does not support the dominated option. But then as we saw at sections 4.2–4.5, neither does EDT support that option in these cases. On the contrary, the presence of an inclination or other screen makes the agent's options evidentially irrelevant to the event, and so entails the unique EDT-rationality of the dominant option. So the evidentialist should be comfortable with this point. Her position will continue to be that *WAR* supports EDT over CDT, because it mandates the dominated option in just *those sorts* of at least apparent Newcomb cases where EDT does the same.

What about *Yankees*? Whether it commits this fallacy depends on what 'now' in (7.22) is supposed to mean. Consider first any moment *after* Mary has learnt whether her next bet will win or lose but *before* she has decided how to bet. It would be fallacious for her then to apply *Yankees* to herself, because it would be fallacious for her then to apply (7.20) to herself. At any such moment she has relevant information putting her in a narrower sub-population than that over which (7.20) generalizes. It puts her not only in the population of bettors but in the sub-population of *winning* bettors, if she has just learnt that she will win, or in the sub-population of *losing* bettors, if she has just learnt that she will lose.

Thus suppose the predictor has just said: 'Mary, you will win your next bet.' Then the statistical generalization that she should apply to herself is not the one that compares the expected return to placing a bet on the Red Sox with the expected return to placing a bet on the Yankees. It is the one that compares the expected return to placing a *winning* bet on the Red Sox with the expected return to placing a *winning* bet on the Yankees. Now we know from Table 7.3 that the expected return to placing a winning bet on the Red Sox is 2 and the expected return to placing a wining bet on the Yankees is 1. Hence the appropriate generalization is not (7.20) but:

(7.30)   The expected return to placing a winning bet on the Red Sox exceeds the expected return to placing a winning bet on the Yankees.

Inferring (7.22) from premises including (7.20) rather than its opposite from ones including (7.30) is just the same fallacy as that of *Why Ain'cha Well*: the fallacy of ignoring available and relevant evidence. So if 'now' in (7.22) refers to a time *after* Mary learns that she will win her next bet, then *Yankees* is invalid.

With appropriate adjustments the foregoing argument will apply if 'now' in (7.22) refers to any time at which Mary has just learnt that she will *lose* her next bet. Hence it is fallacious to apply *Yankees* to Mary once she has learnt the outcome of her next bet, *whatever* she has learnt.[41]

---

[41] Someone might object that conditionalizing on the information that, say, this bet is going to win does nothing to affect Mary's confidence that in the long run and taken over all bets, bets on the Yankees will do better than bets on the Red Sox. So even if she learns that she will win her next bet, is she not still entitled to be as confident in (7.20) as she was before? And in that case, doesn't Arntzenius's argument still go through? But the point is not that Mary's information makes (7.20) false but that (7.20) no longer entails anything about what she should now do. Certainly her next bet belongs to a population of bets of which (7.20) is true. But the oracle's predication also puts it in a *narrower* population of which (7.30) is true. Mary should be applying the generalization about the *narrower* population to her present bet rather than the (equally true) generalization about the

What about the time just *before* Mary has learnt the outcome of her next bet? At those times she does not *have* the evidence that is supposed to vitiate the step to (7.22). So isn't the argument then just as plausible as *Why Ain'cha Rich*?

It's true that it doesn't then commit the *same* fallacy as *Why Ain'cha Well*. The trouble is that now we cannot infer (7.23) from (7.12)–(7.15) and (7.22), because now it no longer follows from these premises that EDT recommends betting on the Red Sox. *Before* Mary learns whether she will win her bet, the news values of betting on the Red Sox and on the Yankees are:

(7.31) $\quad V(BR) = 2\,Cr\,(R \mid BR) - Cr\,(Y \mid BR) = 2.10\% + -1.90\% = -0.7$

(7.32) $\quad V(BY) = -2\,Cr\,(R \mid BY) + Cr\,(Y \mid BY) = -2.10\% + 1.90\% =$
$\quad\quad 0.7$

So at this time EDT recommends betting on the *Yankees*, so once again its preferred option is the one that foreseeably does better.

Yankees is therefore unsustainable for reasons that do nothing to undercut *WAR*. Neither after nor before Mary has learnt whether her bet is a winner does *Yankees* support an option that diverges from EDT in the way that *WAR* supports an option that diverges from CDT. Not after, because then it doesn't *support* a divergent option. Not before, because then it doesn't support a *divergent* option. And this restores the disparity between EDT and CDT. *WAR* does not cut both ways. *It* tells against CDT but not EDT, and no parallel argument tells against EDT but not CDT.[42]

---

broader population. Otherwise, it would be rational not to visit the doctor, even given these rather serious symptoms, on the basis that in the *general* population, people who visit doctors fall sick more often than those who do not.

[42] A similar point applies to the similar objection to *WAR* that Gibbard and Harper (1978: 369–70) based on a variant of the Newcomb case in which *both* boxes are transparent. 'Most subjects find nothing in the first [i.e. what used to be the opaque] box and then take the contents of the second box. Of the million subjects tested, 1% have found 1 million dollars in the first box, and strangely enough only 1% of those... have gone on to take the 1,000 dollars they could each see in the second box.' Everyone should agree (certainly EDT and CDT agree) that it would be rational to take both boxes in this case, whatever you see in them. But according to Gibbard and Harper (1978: 371), *WAR* recommends only taking one box in this modified version of the Newcomb problem. But if it goes wrong here – the argument runs – then there must have been something wrong with it as applied to the standard Newcomb case. But that is not so: the reason it goes wrong here is that the agent has information that puts him not only in the class of *people who face this case*, within which indeed those who take just one box do better on average, but also within the narrower class of *people who face this case and can see 1 milion dollars in the first box*, if that is what he sees. Within that narrower class, the expected return to two-boxing exceeds that to one-boxing by 1,000 dollars. Similarly if the agent is in the narrower class of people who can already see that the first box is empty. But the fact that nothing like *WAR* applies in this modified Newcomb case doesn't stop it applying to the standard Newcomb case, in which no more specific information makes the statistically superior *overall* performance of one-boxing irrelevant to you.

So this response to *WAR* fails. And what remains of the earlier response to it seems to be the intuition, to which section 7.4.3 returns, that becoming a millionaire one-boxer who could have done better is worse than becoming a 'thousandaire' two-boxer who could not. As I said, this is implausible if all that concerns you is your actual wealth, and for this reason *WAR* seems to me to be a strong argument for one-boxing in the standard Newcomb case. But there are also arguments on the other side, which I now examine.[43]

7.4   The case for two-boxing

What makes the standard Newcomb case *seem* paradoxical is that there are apparently strong arguments on either side. One possible reaction to this is that the case itself is impossible, not because of the physical impossibility of a predictor but because it is somehow incoherent. The main alternative is to show that the arguments on one side are not as strong as they seem.[44]

---

[43] Two other objections to *WAR* deserve brief comment. The first is that all it shows is that riches in the Newcomb case are reserved for the irrational. This objection surely begs the question, since what counts as practical rationality is exactly what is at issue between EDT and CDT. (Lewis 1981c seems to acknowledge this.) In fact the objection could be made on behalf of *any* decision rule against *any* demonstration that it either actually does or counterfactually would fare worse than some rival, e.g.: 'All that the dominance argument (for two-boxing) shows is that the predictor reserves a 1,000-dollar bonus for the irrational.' This reveals its vacuity. The second objection is that in a Newcomb problem, the *chance*, at the time of your decision, that the predictor has predicted that you will take the opaque box, is either 1 or 0. So you should have confidence 1 that in an indefinitely extended sequence of repetitions of this problem, two-boxing will on average get you 1,000 dollars *more* than one-boxing. So considerations of what we expect to happen 'in the long run' do nothing to support EDT (Skyrms 1984b: 87–90). But two-boxing only does better than one-boxing when its returns are averaged over a population in which the chance of S_1 is held *fixed at its actual value on the present occasion*. What non-question-begging justification is available for taking *that* sub-population to be of significance? You might as well preach fatalism in *Sink or Swim* (Table 1.4) on the basis that over indefinite repetitions of the case, not learning to swim has a better average return than learning to swim if we hold fixed the actual value of the *outcome* on this occasion. In *WAR* itself, on the other hand, (7.3) and (7.4) are got by averaging over all Newcomb situations, holding fixed only *known* values of relevant variables. There is nothing arbitrary about this if you care about actual wealth over metaphysical money. It reflects, e.g., how the market, or any rational investor, would calculate the expected dividends on shares in a firm that made money by one-boxing, or a firm that made money by two-boxing (Seidenfeld 1984: 204–5).

[44] A third possible reaction is that the standard presentation of the problem *under*specifies it. Once some crucial further question about the case is settled one way or the other, we no longer feel pressure in both directions. For instance, according to Wolpert and Benford (2013: 1640 with trivial changes of notation), the agent in Newcomb's problem should maximize the quantity $F =_{\text{def.}} M\, Pr\, (O_1 \wedge S_1) + (M + K)\, Pr\, (O_2 \wedge S_1) + K\, Pr\, (O_2 \wedge S_2)$ where 'Pr' is some probability function that they do not interpret. They then argue that it is underspecified whether: (a) the agent chooses $Pr\, (O_1)$, but $Pr\, (S_1 \mid O_1)$ is out of his control; or (b) the agent chooses $Pr\, (O_1 \mid S_1)$, but $Pr\, (S_1)$ is out of his control. But *once* we have stipulated which of these is true the paradox is in any case resolved: in case (a) he should unambiguously one-box but in case (b) he should unambiguously two-box (ibid. 1641–2). But this solution is itself underspecified until we know what 'Pr' means. If Pr is the

The first reaction would favour my overall contention that the Newcomb case gives us no reason to prefer CDT to EDT. But nobody has yet vindicated it; and, given that there seems to be at least one relatively realistic model of it, viz the *Prisoners' Dilemma* model at section 4.6, it seems unlikely that the case involves some concealed incoherence.[45]

The second reaction is the one that the ensuing sections pursue. In particular I argue that the arguments for two-boxing should be less convincing than they have been. I'll cover three main lines of thought.

7.4.1 Dominance

The first and most obvious argument for taking the money is a version of the 'dominance' argument that I first stated at section 1.6. You are better off taking the extra 1,000 dollars when there is 1 million dollars in the opaque box. You are better off taking the extra 1,000 dollars when there is nothing in the opaque box. So you had better take the extra 1,000 dollars.

As just stated this argument is invalid. If it were valid then fatalism would be true, since every version of the fatalist argument shares this form. Returning to the case that section 1.6 discussed (*Sink or Swim*: Table 1.4): Alice is better off not having learnt to swim before drowning than having learnt to swim before drowning; she is better off not having learnt to swim and not drowning than she is having learnt to swim and not drowning; so she is better off not learning to swim.

On the other hand there *are* instances of dominance-style reasoning that seem compelling. Returning to *Savage 2.7* (section 1.3, Table 1.2): an investor is better off in farmland than in gold under a Republican president; he is likewise better off in farmland under a Democratic president; so he had better invest in farmland. This reasoning seems correct, but it has the

---

agent's subjective belief function Cr then (a)'s truth and (b)'s falsity are implicit in the standard description of the problem. But this does nothing to settle things in favour of one-boxing, because now the causalist will deny that F is what the agent should be maximizing. On the other hand, if Pr is the objective chance function then the standard description seems to verify both (a) *and* (b). So on neither interpretation of Pr does Wolpert and Benford's analysis resolve the dispute. And if Pr is left uninterpreted then it is not saying anything at all. ₊

[45] Perhaps the outstanding version of this approach derives from the Menzies/Price agency theory discussed at section 5.3.1, according to which the deliberator cannot consistently think that his present choice whether to take the transparent box is both evidentially relevant and causally irrelevant to the prediction. As I indicated there, that line has the consequence that a deliberator should take any contemplated option to be causally relevant to *every* total past cause of her present choice, thus generating a multitude of causal loops that appears both bizarre and ad hoc. But in any case this approach only supports the view that specifically causal information is relevant to practical decision-making if we so liberalize the notion of causal relevance as to drain causalism itself of any interesting content.

same form as the fatalist argument. We should ask what type of additional premise makes dominance arguments valid where they should be, but of which no true instance validates the fatalist argument.

As I have already mentioned, there are two candidates for this, corresponding to the two interpretations of 'independence' in Savage's framework. (a) Dominance arguments require the additional premise that the event is *evidentially* independent of the act. For instance, whether or not the businessman purchases the property tells him nothing about whose candidate will win the election. But Alice's learning to swim *is* an additional reason for her to expect that she will not drown. (b) Alternatively, the additional premise asserts the causal independence of the event and the act. Whether or not the businessman purchases the property has no effect on the election, but Alice's learning to swim *does* have an effect on whether she drowns.

For purposes of evaluating the dominance argument as applied to the Newcomb problem, it is crucial whether it is (a)-type premises or (b)-type premises that make the difference between good dominance arguments and bad dominance arguments. That is because in the Newcomb problem, the relevant (a)-type premise is false whereas the (b)-type premise is true, since whether or not you take the 1,000 dollars in the transparent box is evidentially but not causally relevant to the contents of the opaque one. Unfortunately we cannot settle this question by simply comparing everyday examples of good dominance arguments with fatalistic ones, the reason being that *both* types of premise have relevant true instances when in ordinary life we find dominance arguments convincing, and *neither* type of premise has a relevant true instance in connection with fatalistic reasoning.

So far, then, the position is as follows. The whole question whether the dominance argument applies in Newcomb's problem is the question whether the additional assumption that a dominance argument needs is (a) the evidential or (b) the causal independence of the event on the act. But this is itself something over which EDT and CDT disagree. Moreover, we cannot settle it by appeal to the everyday uses of dominance arguments that both parties accept, because in those cases *both* assumptions are available. So the position over this argument is a stand-off.

What *would* throw light on things would be a sort of converse to the Newcomb case, in which we have evidential independence but *not* causal independence of the events on the acts. If the dominance argument *is* plausible here then that would be grounds to think that it is evidential, not causal, independence that is doing the work when both are present.

Table 7.5 *Armour*

	S_1: die tomorrow	S_2: survive tomorrow
O_1: you get drunk	−2	1
O_2: you buy armour	−3	0

It is difficult to construct a realistic case that combines this pattern of dependence and independence, in which it is intuitively clear what to do. The following case is not realistic but it is intuitively clear, and so deserves at least as much weight as the standard Newcomb case.

> *Armour:* You are a medieval soldier and tomorrow is the day of the big battle. You have the option to buy a suit of armour for thirty florins. But the predictor tells you that you will be *dead* by the end of tomorrow. The predictor's strike rate for this sort of prediction is 99 per cent, whether or not the predictee wears armour. Should you spend what is probably your last thirty florins on a new suit of armour or on getting drunk?

We may represent your options, the possible events and your pay-offs in Table 7.5.

These events are evidentially independent of your act. How you spend the money makes no difference to your 99 per cent confidence that you will die tomorrow. On the other hand, you hardly think that the laws of nature are going to be suspended tomorrow. So of course you think that wearing armour in the battle would increase your *chances* of survival. So here we have evidential independence without causal independence.

Does the dominance argument apply? I think it does. 'You are better off dying and drinking than dying in armour; you are better off surviving and drinking than surviving in armour; and 99 per cent of people who face this prognosis will die, whether or not they wear the armour. So you may as well spend the money on getting drunk.' It simply cuts no ice that wearing armour *causally* promotes survival in battle. To reject the dominance argument for that reason is to ignore information that you have every reason *not* to ignore – and, if it is relevant, that you will regret having ignored as you lie dying in your armour.

Could the causalist argue that in this case we have *causal* as well as evidential independence, because if you are almost certain that you will die then you are almost certain that purchasing armour is causally irrelevant to your survival? No. It remains true that purchasing armour increases your present *chance* of survival; that is, it increases the present value of that physical parameter. Thus if you learn that you will die then you will perhaps reduce your current estimate of the current chance that you *will* wear armour; but you should still reckon the current chance that you will die, *conditional on* wearing armour, to be less than the current chance that you will die, conditional on getting drunk. Arguing that the oracle's prediction makes armour causally irrelevant to survival would be as fallacious as arguing that in Newcomb's problem, your near-certainty that the predictor is right implies near-certainty that two-boxing is causally irrelevant to whether the predictor is right. So we have evidential independence without causal independence. And the dominance argument, which here advocates not wearing the armour, looks like getting it right.[46]

Similarly with other such cases. If the oracle predicts that your wallet will be stolen, the thing to do is to put your cash and credit cards somewhere else, not to conceal your wallet.[47] More generally, myth, literature and popular fiction are filled with stories in which some character, on learning of a prophecy, mistakenly attempts either to help it along (Croesus, *Macbeth*) or to obstruct it (*Oedipus Rex*, 'Appointment in Samarra', 'The Sleeping Beauty'), when he would have been better advised to take for granted what he knew was actually going to happen. 'If chance will make me king, why, chance may crown me without my stir.'

The fact that the case is fanciful is not relevant. The point of the case is not to serve as a direct counterexample to CDT but simply to separate the roles of evidential and causal independence in driving the plausibility of dominance arguments. For that purpose, its being fanciful makes no difference as long as it elicits clear intuitions.

Armour therefore seems to me to be evidence that what makes dominance arguments apply is not the causal but the evidential independence of the events upon the acts. If so, then we also have evidence that the dominance argument for two-boxing in the standard Newcomb problem, which violates the condition of evidential but not of causal independence,

---

[46] It is arguable that Causal Decision Theory itself says nothing about this case, at least not in the present formulation of it. For that formulation relies on the Principal Principle (2.26), which only applies when the agent has no inadmissible evidence; but it may be that the oracle's prediction *is* inadmissible, because it concerns the outcome of the chance process leading to your death, or your survival, in the battle. Still, that is irrelevant to the point that the example is supposed to illustrate, which is only that the principle of dominance plausibly applies to it.

[47] For similar cases see Egan 2007: 100–1.

simply extrapolates in the wrong direction from everyday applications of that style of argumentation.[48]

Of course the point is hardly decisive. Although *Armour* makes a case that evidential independence, and not causal independence, is what is doing the work in dominance arguments, someone might insist that things are not so obvious even there. But even then, things remain deadlocked. The dominance argument cannot show that EDT gets it wrong in the standard Newcomb problem without the use of additional premises that only a causalist has antecedent reason to accept. So at least as far as this argument can show, Newcomb's problem does not constitute grounds for rejecting evidentialism.

7.4.2   Information dominance

A related but distinct argument for two-boxing imagines that your accomplice has seen what the predictor put in the opaque box and is willing to tell you what it is, perhaps for a fee not exceeding 1,000 dollars. If on receiving the fee he tells you that the predictor put 1 million dollars in the opaque box, your best response is clearly to take both boxes, and both EDT and CDT will recommend that. And if he tells you that the predictor put *nothing* in the opaque box, your best response is again to take both boxes, as EDT and CDT again agree. So why bother paying the fee? If you know that *whatever* extra information you are about to get, it will, on getting it, be rational for you to two-box, isn't it already rational to two-box *before* you get it? In fact, isn't it rational to do so even if there *is* no accomplice?[49]

The answer is that it doesn't follow, from the premise that two-boxing is rational for an agent who knows what is in the opaque box, that it is also rational for an agent who does not yet know this, let alone for one who is not in any position to know this until after he has chosen. Before one gets the information and after one gets it one is facing *different problems*. Before getting the information the agent takes her acts to be symptomatic of what

---

[48] An alternative response, consistent with this stance on *Armour*, might be that neither causal nor evidential dependence is necessary for the dominance argument to apply, but either by itself is sufficient. On that view you should take both boxes in the standard Newcomb problem *and* get drunk in *Armour*. That line seems arbitrary and unmotivated. Depending on what view one took of cases involving neither sort of dominance, one would have to say that either the evidential or the causal bearing of one's acts is irrelevant *except* in cases of dominance relative to an evidentially or causally independent partition of *S*, when suddenly and for no apparent reason it becomes decisive.

[49] This summarizes an argument due to Pollock (2010: 70–3). There is a rudimentary version of it in Nozick (1969: 209–10).

is in the box; *after* getting it she does not. So there is nothing wrong with taking different acts to be rational *ex post* and *ex ante*. At least that is so unless changes in the symptomatic bearing of one's acts on the relevant events are practically irrelevant if unaccompanied by changes in what one knows about their causal bearing. But to assume that at the outset would be to beg the question against EDT.

Thus reconsider Alice in *Sink or Swim* (Table 1.4) and suppose that a reliable oracle is about to tell her whether she is going to drown. She reflects (i) that if the oracle tells her that she will drown, she is better off not learning to swim. And (ii) if the oracle tells her that she will *not* drown, she is better off not learning to swim. Unfortunately the oracle is called away on urgent business and cancels her appointment with Alice. It would not be reasonable for Alice now to think that since she would be better off not learning to swim whatever the oracle was about to tell her, she is in fact already rational not to learn to swim.

Of course the causalist might reject Alice's initial reflections (i) and (ii). He may deny, as in *Armour* at section 7.4.1, that once she knows what will happen to her she is rational not to learn to swim. But these reflections are intuitively plausible and so establish the intuitive invalidity of this form of argument, which therefore cannot be used to justify two-boxing to anyone who does not already accept CDT.[50]

That argument for two-boxing might seem attractive to anyone who is attracted to Arntzenius's **desire reflection** principle, that one should not 'be such that one can foresee that one's future desires will differ from one's current ones in such a way that one will later regret them'.[51] This principle *does* prohibit taking different conative attitudes towards one-boxing *ex post* and *ex ante*. But why believe it is true? It is an instance of the Piaf maxim discussed at section 3.2. But as we saw there, that principle is itself dubious.

Finally, one might think that quite independently of whether they issue in foreseeably regrettable actions, there is something wrong in principle with the idea that one's preferences, as reflected in relative news values, should be such that one knows in advance that they will be reversed in the face of information that one is about to receive. If you know *that* in advance, shouldn't you reverse them in advance?

No: foreseeable fluctuations in preference may be perfectly rational, even in cases where the fluctuation is in a known direction and is a rational response to future information.

---

[50] For a different and more CDT-sympathetic counterexample to the argument that I am here attacking, see Joyce 2007: 550–1.

[51] Arntzenius 2008: 278.

The following example illustrates this. The admissions statistics for the English and Mathematics Departments at Simpson's Paradox University (SPU) are what you would expect. Male applicants are less successful than female ones overall: 14 per cent of men who apply for admission to a graduate course in one of these departments are successful, but 20 per cent of women who so apply are successful. But in each department the discrepancy is reversed: 5 per cent of male applicants for Mathematics are successful as against 1 per cent of female applicants. And 50 per cent of male applicants for English are successful as against 25 per cent of female applicants. The explanation is that male candidates are more likely than female candidates to apply to the more competitive Mathematics Department.

Your best friend has applied to graduate school at SPU. It matters greatly to you that the application is successful. You know that your friend applied to the English or Mathematics Department, but not to both. But for some reason you have forgotten (a) which it is and (b) whether your friend is male or female. You ask your friend about (a).

Before hearing the answer you reflect that *now*, you'd prefer the answer to (b) to be that your friend is female. After all, female applicants to SPU do better than male ones. You then reflect that *after* hearing the answer to (a) and *whatever it is*, the news value of the information that your friend is male will *exceed* the news value of the information that your friend is female. After all, male applicants to Mathematics do better than female ones, and male applicants to English do better than female ones. Finally, you reflect that now, before getting the answer to (a), you have a preference over the possible answers to (b) that you *know in advance* is going to be reversed in the light of the answer to (a).

But all this is quite rational. At any point in time you have just the preference that is appropriate in the light of all of the information that you then possess. The only *peculiarity* of the situation is that the direction of change is foreseeable. But that change is itself a perfectly rational response to a commonplace statistical phenomenon.

7.4.3 'Could have done better'

The third argument for two-boxing is really another kind of dominance consideration. But it is worth considering separately, because (a) it is intuitively a different formulation of the thought behind dominance; and (b) it picks up on Joyce's response to *WAR* as covered at section 7.3.2.

Table 7.6 *Newcomb-Insurance* stage 1: dollar
pay-offs

	S_1: predicts O_1	S_2: predicts O_2
O_1: take opaque box	100	0
O_2: take both boxes	100	0

We might say in defence of two-boxing that the two-boxer is doing *as well as she could* be doing in the circumstances. Given the contents of the opaque box she could not have done better than to take both boxes. If she had taken just one box then she would have been 1,000 dollars worse off.

More precisely: define your present *situation* as the totality of facts that you cannot now causally affect. For instance, in the standard Newcomb problem the contents of the opaque box are, at the moment of choosing, part of your situation. Then the reason for two-boxing is that only it conforms to the following principle:

(7.33)   **CDB**: If you know that a certain available option makes you worse off, given your situation, than you would have been on some identifiable alternative, then that first option is irrational.[52]

'CDB' abbreviates 'Could have done better'.

Many philosophers have taken CDB as decisive grounds for two-boxing and not one-boxing.[53] But this argument is not available to the causalist. For as I'll argue, CDT itself violates the principle. To see this, consider the following *Newcomb-Insurance* problem.

The problem is in two stages. Stage 1 is a degenerate Newcomb problem in which the predictor, whose strike rate is $n > 0.75$, puts $100 in an opaque box if and only if he predicted that you wouldn't also take the visibly *empty* box that is adjacent to it. (This degeneracy is not essential for the example to work, but it simplifies the arithmetic.) Your dollar pay-offs at stage 1 are therefore as in Table 7.6. Stage 2 occurs immediately after stage 1, so before

---

[52] The point of 'identifiable' is that not betting on either side of a causally independent proposition will certainly make you worse off than you would have been in your actual situation had you taken *one* of those two bets, only you don't know which. And yet it may still be rational for you not to bet.

[53] E.g. Lewis (1979a: 303–4, 1981a: 309–12); Skyrms (1984b: 67); Joyce (1999: 152–4); Weirich (2001: 126); Sloman (2005: 94).

Table 7.7 *Newcomb-Insurance* stage 2: dollar pay-offs

	T_1: predictor right	T_2: predictor wrong
P_1: bet predictor right	25	−75
P_2: bet predictor wrong	−25	75

Table 7.8 Net pay-offs in
Newcomb-Insurance problem

	S_1: predicts O_1	S_2: predicts O_2
$O_1 \wedge P_1$	125	−75
$O_1 \wedge P_2$	75	75
$O_2 \wedge P_1$	25	25
$O_2 \wedge P_2$	175	−25

you get to open the opaque box. You must bet for or against the predictor's having *correctly* predicted you at stage 1. *Either* you bet seventy-five dollars at odds of 1:3 that he got it right, *or* you bet twenty-five dollars at odds of 3:1 that he got it wrong. So your dollar pay-offs at this stage are as in Table 7.7. The predictor makes no prediction about stage 2, of which he neither knows nor cares.

Finally all is revealed. You learn what is in the opaque box. You learn whether the predictor got you right at stage 1. You learn whether you have won whichever bet you took at stage 2. So you also learn your *net* pay-off, which depends on your choices and the predictor's prediction as stated in Table 7.8, which amalgamates Tables 7.6 and 7.7. For instance, row 3 of Table 7.8 corresponds to your taking both boxes and then betting that the predictor predicted this. If the predictor did *not* predict this but instead that you'd one-box, you win $100 at stage 1, because that is what he put in the opaque box, but lose seventy-five dollars at stage 2, because you wrongly bet on his getting it right, for a net return of twenty-five dollars. If the predictor *did* predict that you'd take both boxes at stage 1 then you win nothing from the opaque box but twenty-five dollars at stage 2, again for a net return of twenty-five dollars. It is this smoothing of pay-offs in rows 2 and 3 that motivates my calling it an insurance problem.

Table 7.8 is misleading in one respect. It makes the problem seem to involve a *single* decision between *four* options. In fact it is a genuine case of *sequential* choice. You face two successive decisions, each between two options. Table 7.8 simply summarizes net returns to each overall course of action.

What happens if you let CDT guide you through the sequence? That depends on how you approach problems of sequential choice in general. I will discuss two possible such approaches: (a) the 'myopic' approach and (b) the 'sophisticated' approach. (There is a third, 'resolute', approach, which I omit here but briefly discuss in n. 59 and the accompanying main text.)

(a) *Myopic choice.* The myopic approach towards sequential choice involves selecting, at each stage, the overall sequence that then seems best to you out of those still available, and then realizing the presently available component of that best sequence.[54] In the present case it applies as follows: at stage 1, you decide which of the four possible sequences is best according to your favoured decision theory. Then you realize the first step of that sequence. At stage 2, you decide again on the same basis between the two sequences that are still possible. Then you realize the second step of the sequence that seems best to you then.

Writing $A \succ_n B$ for 'A is strictly more choiceworthy than B for you at stage n', suppose just for illustration's sake that $O_1 \wedge P_1 \succ_1 O_1 \wedge P_2 \succ_1 O_2 \wedge P_1 \succ_1 O_2 \wedge P_2$ and that at stage 2 this ranking is exactly reversed. Then at stage 1 you will, if you choose myopically, realize O_1. And at stage 2 you will realize P_2, because $O_1 \wedge P_2 \succ_2 O_1 \wedge P_1$, and these are the only sequences still available. So you will end up with $O_1 \wedge P_2$.

Whether you take a myopic attitude towards sequential choice is independent of whether you follow Causal Decision Theory or Evidential Decision Theory. The role of CDT and EDT is to decide which of the sequences is choiceworthy, out of the ones that then remain available to you. You will use whichever of CDT and EDT you follow to answer this question as it arises at each stage.

What does CDT recommend at each stage? At stage 1 there are four sequences available to you, as summarized in Table 7.8. The pay-off to each option depends on what the predictor has predicted. From the CDT perspective we can immediately rule out the sequences $O_1 \wedge P_1$ and $O_2 \wedge P_1$, since both are dominated: $O_2 \wedge P_2$ always beats $O_1 \wedge P_1$ by fifty dollars, and $O_1 \wedge P_2$ always beats $O_2 \wedge P_1$ by fifty dollars, and the prediction is

---

[54] Strotz 1955: 168–71.

causally independent of anything that you now do. For the remaining two sequences we have:

(7.34) $\quad U_{\scriptscriptstyle \rm I} (O_{\scriptscriptstyle \rm I} \wedge P_{\scriptscriptstyle 2}) \geq U_{\scriptscriptstyle \rm I} (O_{\scriptscriptstyle 2} \wedge P_{\scriptscriptstyle 2})$ iff $75 \geq 175 Cr_{\scriptscriptstyle \rm I} (S_{\scriptscriptstyle \rm I}) - 25 Cr_{\scriptscriptstyle \rm I} (S_{\scriptscriptstyle 2})$

Here the subscripts '1' after U and Cr indicate that what we are discussing are your utilities and credences at stage 1.

It follows that at stage 1 CDT endorses $O_{\scriptscriptstyle \rm I} \wedge P_{\scriptscriptstyle 2}$ to the myopic agent if $Cr (S_{\scriptscriptstyle \rm I}) \leq 0.5$. If $Cr (S_{\scriptscriptstyle \rm I}) \geq 0.5$ it endorses $O_{\scriptscriptstyle 2} \wedge P_{\scriptscriptstyle 2}$. So if the myopic causalist has $Cr (S_{\scriptscriptstyle \rm I}) < 0.5$ at stage 1 then he will take just one box at stage 1; if he has $Cr (S_{\scriptscriptstyle \rm I}) > 0.5$ at stage 1 then he will take both boxes at stage 1; and if $Cr (S_{\scriptscriptstyle \rm I}) = 0.5$ then he may do either of these things.

At stage 2, you have a choice between two sequences. Which sequences these are will depend on what you chose at stage 1. If you chose $O_{\scriptscriptstyle \rm I}$ at stage 1 then your choice is between $O_{\scriptscriptstyle \rm I} \wedge P_{\scriptscriptstyle \rm I}$ and $O_{\scriptscriptstyle \rm I} \wedge P_{\scriptscriptstyle 2}$. If you chose $O_{\scriptscriptstyle 2}$ at stage 2 then you now have a choice between $O_{\scriptscriptstyle 2} \wedge P_{\scriptscriptstyle \rm I}$ and $O_{\scriptscriptstyle 2} \wedge P_{\scriptscriptstyle 2}$.

But on the question whether or not to bet at stage 2, it just doesn't matter what you did at stage 1. It doesn't even matter whether you remember what you did. Whatever you did, the right option now is to bet on the predictor's having been *right*, i.e. to take $P_{\scriptscriptstyle \rm I}$. After all, your choice has no bearing, causal *or* symptomatic, on what it is a bet upon. And you have a choice between betting $(P_{\scriptscriptstyle \rm I})$ at 1:3 *on*, or $(P_{\scriptscriptstyle 2})$ at 3:1 *against*, a proposition in which your present confidence $= n$ exceeds 0.75. S. Ulam once defined a coward as someone who wouldn't take odds of 2:1 on *either* side of a proposition.[55] But if n greatly exceeds 0.75 then even a coward would gladly take the bet that $P_{\scriptscriptstyle \rm I}$ represents.

And that is what CDT advises. Consider Table 7.8 and suppose that you have, and know that you have, two-boxed at stage 1. Then your options are effectively the bottom two row-headings in Table 7.8, i.e. $O_{\scriptscriptstyle 2} \wedge P_{\scriptscriptstyle \rm I}$ and $O_{\scriptscriptstyle 2} \wedge P_{\scriptscriptstyle 2}$. Which one you realize *makes* no difference either to the content or to the correctness of the prediction. But since you know that you chose $O_{\scriptscriptstyle 2}$, your confidence that he predicted this is $n > 0.75$. So the stage 2 U-scores of the available options are:

(7.35) $\quad U_{\scriptscriptstyle 2} (O_{\scriptscriptstyle 2} \wedge P_{\scriptscriptstyle \rm I}) = 25$
(7.36) $\quad U_{\scriptscriptstyle 2} (O_{\scriptscriptstyle 2} \wedge P_{\scriptscriptstyle 2}) = 175(1 - n) - 25n = 175 - 200n$

Since $n > 0.75$ we have $U_{\scriptscriptstyle 2} (O_{\scriptscriptstyle 2} \wedge P_{\scriptscriptstyle \rm I}) > U_{\scriptscriptstyle 2} (O_{\scriptscriptstyle 2} \wedge P_{\scriptscriptstyle 2})$. A similar argument proves that $U_{\scriptscriptstyle 2} (O_{\scriptscriptstyle \rm I} \wedge P_{\scriptscriptstyle \rm I}) > U_{\scriptscriptstyle 2} (O_{\scriptscriptstyle \rm I} \wedge P_{\scriptscriptstyle 2})$. Hence whatever you did at

---

[55] Samuelson 1963: 98.

stage 1, CDT endorses P_1 at stage 2. So in fact does EDT, and so too does common sense.

The upshot of all of this is as follows.

(7.37)   The myopic agent who follows CDT will choose the sequence $O_1 \wedge P_1$ if $Cr_1(S_1) < 0.5$ and $O_2 \wedge P_1$ if $Cr_1(S_1) > 0.5$. If $Cr_1(S_1) = 0.5$ then he may choose either.

(b) Sophisticated choice. What makes 'myopic' appropriate is that the myopic agent will initiate or continue a plan *when* it seems ideal to him, without seeing ahead, i.e. without caring whether he will follow through on its later stages. This *looks* irrational, although I won't argue here that it is. In any case, the alternative *sophisticated* approach to sequential choice differs from myopic choice on just this point. The sophisticated approach to sequential choice is to work backwards from the final stage. If you can now foresee that by your lights at that final stage a certain option will seem rationally compelling, then you should now take it for granted that you will realize that option. (If you cannot, because your choice at that stage causally depends on what you do at previous stages, then you should evaluate the options at those previous stages taking these causal effects into account. As we shall see, that kind of indeterminacy doesn't arise in this problem.) Then you should evaluate what will seem rational at the penultimate stage, *given* your already predetermined decision at the final stage. Iterate until you reach the present stage.

The sophisticated approach is probably the most popular approach to sequential choice. Certainly, those few philosophers who have tried to apply CDT to problems of sequential choice have endorsed it implicitly.[56]

---

[56] For instance Maher (1990: 482–4) argues that CDT should evaluate the immediately available options (not sequences) at the current stage by taking into account their effects on one's choices at later stages. So if you know that you are going to realize some option p^* at stage 2 of a two-stage decision problem, whatever you do now, then you should take your stage 1 choice to be in effect a selection from amongst $\{o \wedge p^* \mid o \in O\}$, where O is the set of your stage 1 options. This is what the sophisticated approach recommends in *Newcomb-Insurance*. Joyce (1999: 60–1) takes the immediately available options at any stage in sequential choice to be *resolutions*: whether or not your present self can influence your future choices, it is entirely up to your present self what sequence it now *resolves* to actualize. But again, when CDT chooses from amongst resolutions, it should take into account the causal effect that a present resolution to realize (say) p^* at stage 2 will actually have on what you do at that future time. If you know in advance that it has no effect, because you know that at stage 2 you will realize p^* come what may, then you should simply treat your future choice as just another unalterable fact about the state of the world. So Joyce is committed to the sophisticated approach, at least in the case that we are presently discussing. Similarly, Arntzenius, Elga and Hawthorne (2004: 267) argue that if your present choice has no causal influence over your future choice then it is rational to evaluate one's present options whilst keeping one's future

Table 7.9 CDT and the
Newcomb-Insurance problem

	Myopic	Sophisticated
$Cr_1 (S_1) < 0.5$	$O_1 \wedge P_1$	$O_2 \wedge P_1$
$Cr_1 (S_1) = 0.5$	$O_1 \wedge P_1$	$O_1 \wedge P_1$
	$O_2 \wedge P_1$	$O_2 \wedge P_1$
$Cr_1 (S_1) > 0.5$	$O_2 \wedge P_1$	$O_1 \wedge P_1$

In the present situation, you will if you are sophisticated first look *forward* to stage 2 and ask what will then seem best to you by the lights of your favoured decision theory. Once you have reached an answer on that point, you then make your choice at stage 1, given what you now know about what you are going to do at stage 2.

So suppose that you are a sophisticated chooser and start again with the further supposition that you follow CDT. Looking forward to stage 2: the argument covering myopic choice should already make it clear that whatever you choose at stage 1, you *will* certainly bet on the predictor's accuracy at stage 2. Briefly to recap the point: it is still true that your stage 2 beliefs involve $Cr_2 (T_1) > 0.75$, *whatever* you do at stage 1. And certainly your choice of bet at stage 2 is causally irrelevant to whether T_1 is true. After all, T_1 is a biconditional relating two propositions (O_1 and S_1) that describe states and events that obtain or occur strictly before stage 2, and so nothing that you do at stage 2 can make any difference to their truth-value, nor therefore to T_1's truth-value.

This means that at stage 1 you take yourself to be choosing in effect between $O_1 \wedge P_1$ and $O_2 \wedge P_1$. The stage 1 U-scores of these options are:

$$(7.38) \quad U_1 (O_1 \wedge P_1) = 125 \, Cr_1 (S_1) - 75 \, Cr_1 (S_2)$$
$$(7.39) \quad U_1 (O_2 \wedge P_1) = 25$$

So CDT directs the sophisticated agent to take O_1 at stage 1 if $Cr_1 (S_1) > 0.5$, and it directs him to take O_2 at stage 1 if $Cr_1 (S_1) < 0.5$. If $Cr_1 (S_1) = 0.5$ then both options are permissible.

Table 7.9 summarizes the discussion in (a) and (b) above. The table shows that CDT recommends one of two courses in the *Newcomb-Insurance*

actions fixed, just as sophistication requires. It is true that their paper covers countably *infinite* choice sequences. But nothing in their discussion suggests any reason to restrict *this* point to the infinite case.

problem. *Either* it recommends one-boxing and then recommends betting that the predictor was right about that, *or* it recommends *two*-boxing and then recommends betting that the predictor was right about *that*.

In either case it violates CDB. Suppose CDT recommends O_1 and then P_1, perhaps because you are sophisticated and your $Cr_1(S_1) > 0.5$. Table 7.8 shows that whatever your situation – whatever the predictor predicted, this being something that you never could causally affect – $O_1 \wedge P_1$ does worse than $O_2 \wedge P_2$ by fifty dollars. Taking O_1 and then P_1 is throwing away fifty dollars that anyone in your situation could have got if she had taken O_2 and then P_2.

Or suppose CDT endorses O_2 and then P_1, because your $Cr_1(S_1) \leq 0.5$. By Table 7.8 again, the return to $O_1 \wedge P_2$ exceeds the return to $O_2 \wedge P_1$ by fifty dollars whatever your situation. Taking O_2 and then P_1 is just throwing away fifty dollars that you *would* have had if you *had* acted otherwise. So whatever your initial beliefs and whether you are myopic or sophisticated, CDT leaves you fifty dollars worse off than anyone in your situation could have been if they had taken some alternative course that was identifiable as thus superior in advance. So if CDB is true then CDT is false.

It might arouse suspicion that the case involves sequential choice and not a one-off decision. There are four suspicions in particular that it's worth briefly allaying.[57]

---

[57] An approach to sequential choice that I do not mention below, but which certainly is worth considering, is Weirich's very recent suggestion that the agent attaches an intrinsic disvalue to missing opportunities for arbitrage – and equivalently, then, to incurring sure losses relative to some possible course of action (Weirich 2013). From this point of view, the agent who has already taken O_1 at stage 1 has a reason not to take P_1 at stage 2, a reason that would not apply if she were facing only the choice between P_1 and P_2 – for instance, if she were facing a bet on whether the predictor has got it right in a Newcomb scenario in which she was not involved. As we can see from Table 7.8, taking O_1 and then P_1 amounts to throwing away fifty dollars relative to some alternative sequence (in this case, the sequence $O_2 \wedge P_2$). So according to Weirich, her having taken O_1 at stage 1 gives her a reason to take P_2 at stage 2. Weirich's idea deserves more extended discussion than is feasible in this place. But two brief points are appropriate. First, it would surely be possible to *stipulate* that in this example the agent cares *only* about terminal wealth and not at all about 'missed opportunities': she would rather (say) get eleven dollars and miss an opportunity for arbitrage, than get ten dollars in the absence of any such opportunity. Second, whilst it is of course natural to want not to make sure losses, it is implausible that this consideration in fact carries much weight at the outset of stage 2, for an agent who has taken O_1. That agent thinks quite reasonably that the predictor has almost certainly predicted O_1 (i.e. that S_1 is actually true). So she is now facing a choice between (P_1) a bet on a practical certainty and (P_2) a bet against it, the certainty being something on which her bet has neither evidential nor causal bearing. This choice looks completely straightforward. Although it is true that taking the bet would lead to a sure loss relative to a sequence ($O_2 \wedge P_2$) that is no longer available since she has taken O_1, it is unclear that this would actually matter to anyone at stage 2 who was facing the choice that I just described.

(a) If we do think of the problem as a one-off choice between four alternatives, as Table 7.8 misleadingly suggests, then CDT does *not* recommend $O_1 \wedge P_1$ or $O_2 \wedge P_1$. Instead it endorses $O_1 \wedge P_2$ or $O_2 \wedge P_2$, which one depending on (i) the precise value of $Cr(S_1)$ and (ii) whether you are a myopic or a sophisticated agent. You might infer that in the sequential problem, it would be advantageous to be able to *bind* yourself, at stage 1, to taking P_2 at stage 2. But the fact that self-binding can be helpful for followers of some decision rule needn't be grounds to reject that rule.

Of course it's true: the fact that self-binding is the best option *where available* doesn't refute any rule that endorses it. But a sequential problem in which self-binding is available at the outset is a *different problem* from *Newcomb-Insurance*. We know that the latter is *a* decision situation in which CDT violates CDB. The objector is just pointing out *another* decision situation in which it doesn't violate CDB. But that's hardly news. We already knew that CDT doesn't violate the principle in the standard Newcomb problem. What is the further relevance of the self-binding case?[58]

And besides, how *could* it be rational to bind yourself at the outset to P_2? You know at stage 2 that the predictor has got a strike rate of $n > 0.75$. This by itself makes it rational to bet at stage 2 that he got it right this time. But you knew *that* at the outset, and nothing happened to change your confidence that he got it right. So if it's rational at stage 2 to bet that he is right on this occasion, then it must have been rational all along.[59]

(b) Violations of CDB arise in sequences where your choices at one stage affect beliefs that are relevant to what you do at later stages. But these are not problematic.

Consider: you must choose at stage 1 whether to take this legal and performance-enhancing but mildly unpleasant drug. You must choose at stage 2 whether to enter a race that you have every chance of winning if and only if you have taken the drug. But the drug impairs your judgement. If you take it at stage 1 then at stage 2 you'll *think* that you won't win.

---

[58] Meacham (2010: 67) makes a similar point about a similar point.

[59] Similar remarks apply to the closely related *resolute* approach to sequential choice, according to which one both can and should make a plan at stage 1 and stick to it. For exposition and extensive defence see McClennen 1990. My main objection is that resolution is simply not feasible in *Newcomb-Insurance* as I have described it, unless it is rational at stage 2 to stick to the plan that you have made at stage 1. But it isn't: any *plan* that CDT endorses at stage 1 involves P_2 at stage 2, but I can't see how the mere fact of having made this plan *could* make it rational to stick to it at stage 2. Broome 1992 clearly articulates this general pattern of concern about resolute choice.

An otherwise sensible decision rule D might well endorse your taking the drug at stage 1 but then endorse your not entering the race at stage 2, leaving you definitely worse off, whatever your situation, than if you'd either refused the drug or entered the race. But it's arguable that this doesn't refute D. All it reveals, you might think, is that even sensible decision theories go wrong in pathological sequences where your earlier actions distort your later beliefs.

I doubt that, but set that aside. The main point is that the *Newcomb-Insurance* situation is *not* like this. To repeat: at stage 2 you are choosing the direction and stake of a bet on the predictor's accuracy at stage 1. The *only* belief that affects this is your $Cr_1(T_1) = Cr_2(T_1) = n$.[60] This credence is completely independent of what you do at stage 1 and in fact persists throughout the decision problem, as does your knowledge of it.[61]

(c) Analogous suspicions arise about sequential cases in which one's *desires* – that is, preferences over outcomes – change from one stage to the next. It may just be a fact of life that these cases exist. For instance, our preference-bias towards the future (desire for future pleasures; indifference to past pains) arguably leads to foreseeable certain losses for any agent following decision rules that are otherwise sensible. And it is perhaps plausible that in *this* case, all the blame should lie with the temporal bias and not with whatever decision rule was complicit in the result.[62]

Be that as it may, it is irrelevant to the *Newcomb-Insurance* example. Both the pay-off structure and your attitude towards it, i.e. caring only about your terminal wealth, remain constant throughout the sequence. The paradox is that in spite of this fact and your clear-eyed appreciation of it, CDT enjoins throwing away fifty dollars that could certainly have been yours. Any blame for *this* fact rests with CDT alone.

---

[60] But also $Cr_2(S_1 \mid O_1)$ and $Cr_2(S_2 \mid O_2)$ if these exist (e.g. if you can't quite remember what you did at stage 1, so that $Cr_2(O_1)$, $Cr_2(O_2) > 0$), in which case they are also equal to $Cr_1(S_1 \mid O_1) = Cr_1(S_2 \mid O_2) = n$.

[61] You might doubt this, on the grounds that if, for instance, you start out fairly confident that the predictor has predicted that you will take the transparent box (so that $Cr_1(S_1) \leq 0.5$), and then you take only the opaque box, then you will become fairly confident that the predictor is wrong. But that just means that you had no confidence in the predictor's abilities to start with, so we have no Newcomb problem at all. The standard version of the Newcomb problem, which *Newcomb-Insurance* simply appropriates at its stage 1, takes it for granted that *whatever* you choose you will expect that the predictor has predicted you correctly, i.e. your $Cr_1(S_1 \mid O_1)$ and your $Cr_1(S_2 \mid O_2)$ are both high. Otherwise the standard Newcomb problem would itself involve no disagreement between EDT and CDT and so would be powerless to motivate causalism.

[62] Dougherty 2011.

(d) Even if we agree that CDT leads to some dominated sequence in this case, it doesn't follow that there is any tension between it and the CDB principle itself. As stated, CDB concerned the rationality of individual *options*. But (the objector continues) what the example shows is only that CDT violates a different principle, concerning not options but *sequences* of options:

(7.40) **CDB-sequence**: If you know that a certain available *sequence* of choices makes you worse off, given your situation, than you would have been on some identifiable alternative, then that first *sequence* is irrational.

I concede (i) that CDB and CDB-sequence are distinct theses and (ii) that strictly speaking it is only CDB-sequence, not the CDB principle itself, that CDT violates in *Newcomb-Insurance*. But accepting CDB and not CDB-sequence looks completely unmotivated. The basic idea behind both principles is that it's irrational knowingly not to *do the best you can*, given your situation (that is, holding fixed everything that is causally independent of what you do). This idea, which looks as compelling as the narrower CDB itself, would seem to cover options and sequences of options indifferently. Certainly and at least, we are owed a story about why this basic idea is more plausible when applied only to individual options than it is when applied to other courses of action. In the absence of some such story, the causalist cannot appeal to CDB in defence of two-boxing, because CDT itself looks incompatible with the basic idea behind CDB.

My own view is that in the *Newcomb-Insurance* problem you should take O_1 and then P_1. This is in line with EDT. Of course this violates CDB, for in either situation $O_1 \wedge P_1$ does worse than $O_2 \wedge P_2$. But that EDT violates CDB is not news. We already knew that anyone who takes one box in the Newcomb problem could have done better, given his actual situation and whatever that situation is. But the point of this section is that *this* is not a stick with which causalists can beat the one-boxer.

7.5 Conclusion

I have argued that cases of disagreement between EDT and CDT generally do not arise in everyday life (sections 4.1–4.5); that in unlikely but feasible cases where they do arise, there is every reason to find it either unclear that EDT gets it wrong (sections 4.6, 7.2–7.4), or clear enough that EDT gets it right (Chapters 5–6); and that in the standard case that has become central to philosophical discussion of the topic, there are reasons to doubt the case for preferring what CDT recommends (sections 7.3–7.4). If, as argued at Chapter 3, CDT correctly specifies how causal information should enter into rational deliberation if it has a special place there at all, all this supports my main conclusion: that causal information does *not* have any special place in rational deliberation. For practical purposes, a rational agent can base any decision upon a comparison of the news value of his options, quite independently of what he thinks of their causal bearing on the outcomes.

CHAPTER 8

'The ultimate contingency'

This chapter steps back from the details of particular cases over which EDT and CDT disagree and asks whether there is any general philosophical argument for preferring CDT. Why should just *this* relation, the relation of causal dependence, play the role in practical deliberation that CDT mandates? Causalists have generally made little effort to address the question, focusing instead on the particular examples that have been the main focus of chapters 4 and 7.

I know of one important argument to this effect, due to Ramsey. If effective, it shows that a practical concern with the effects of one's choices is actually a de facto consequence of EDT. So if effective, it shows that CDT is a de facto consequence of EDT. More specifically, it says that a rational deliberating agent will take her own contemplated acts as *evidentially irrelevant to any past event*. On plausible assumptions that I am not going to question here, it follows that EDT and CDT will recommend the same options in all of the cases over which I have hitherto taken them to disagree. More specifically it follows that EDT actually recommends just what in all those cases I took CDT alone to recommend: pressing the button in *Psycho Button A*, taking both boxes in the standard Newcomb problem, etc.

By itself, the argument doesn't establish the main thesis that I have been opposing, i.e. the thesis that practical deliberation goes wrong unless it takes account of the agent's specifically causal beliefs. For all that it shows, an agent who has no causal beliefs could still make all the right decisions – or at any rate, decisions that no knowledge about the causal influence of her options would give her reason to revise.

On the other hand, in conjunction with a certain view of causal belief the argument *would* deliver that result. Specifically, if we *identify* an agent's beliefs about the causal efficacy of her options with her beliefs concerning their evidential relevance as evaluated whilst holding fixed the past, then the argument shows that the rational evidentialist is *already* acting on the basis of her causal beliefs.

For instance, suppose one thinks that to hold an event E causally dependent on an event C is (i) to think that C is prior to E; (ii) to hold C evidentially relevant to E holding fixed everything in the *past* of C. Then one takes C to be causally relevant to a subsequent E iff one's credences satisfy:

(8.1) $\quad Cr(E) \neq \Sigma_{X \in P} \, Cr(E|C \wedge X) Cr(X)$

– where P is a partition of the event space into all possible descriptions of what happened before the time of C.[1] Now if C describes a currently available option, and the agent takes C to be evidentially irrelevant to the past, then $Cr(X) = Cr(X|C)$ for any $X \in P$. It follows that (8.1) is true if and only if $Cr(E) \neq Cr(E|C)$. So from this agent's perspective, causal relevance coincides with evidential relevance. So if her beliefs about the latter enter into practical decision-making then so do her beliefs about the former.

And even without that additional premise about causal belief, the argument might still vindicate Causal Decision Theory itself in the weaker sense that it is in fact no worse than EDT, at least for all that I have said here. That is, it would show that even if rational deliberators need not think of the causal efficacy of their options as such when deciding amongst them, they *should* be sensitive to just those relations of evidential dependence that *track* the causal dependency of events upon options, at least on certain views of what the causal relation is supposed to be.

For instance, suppose that we identify causal dependence with chance dependence, so that E is causally dependent on C just in case:

(8.2) $\quad Ch(E|C) \neq Ch(E|\neg C)$

– where Ch represents the chance function at a time just before C is settled.[2] Now suppose that C is a present option for the agent. Then EDT dictates that the agent should evaluate C on the basis of its diagnostic bearing, i.e.

---

[1] This is the subjective analogue of Reichenbach's (1971: 204) theory of causal relevance.

[2] This very simple theory is certainly false if taken to analyse our ordinary concept of causal relevance. Whether C is causally relevant to E may also depend upon chance occurrences that are causally independent of C and take place between the time of C and the time of E: see Kvart 2004 for further details. But it *is* plausible that a causalist should, when evaluating an option C, care only about C's causal relevance to E to the extent that (8.2) captures it. So if the argument in the main text is correct then any evidentialist deliberator whose conditional credence function $Cr(x|C)$ coincides with her expectation of $Ch(x|C)$ will, willy-nilly, end up endorsing the same options as the causalist in Newcomb's problem and in the other cases that were supposed to divide them.

the marginal credence function $Cr_C(x) =_{\text{def.}} Cr(x\,|\,C)$. But if the agent takes the past to be evidentially independent of C, then the conditional version of the Principal Principle (2.26)(ii) demands that the latter should track the marginal chance distribution $Ch_C(x)$ in the sense of equalling the agent's expectation of this quantity.[3] So even an evidentialist agent will aspire to evaluate his options on the basis of what is in fact their causal bearing on the events of interest.

The argument is therefore that a decision theory that is effectively CDT arises quite naturally from an *evidentialist* starting point. According to it, beliefs about the causal bearing of his acts are of special relevance to an agent because they are the beliefs that an evidentialist deliberator can't help acting on, or at least acting as if he is acting on. And the reason for this is supposed to be that the agent takes his present choice to be diagnostically irrelevant to *anything* in its past.

And if this argument is right then it would make many of the preceding examples irrelevant. There simply could not *be* cases where EDT and CDT clash, at least if we set aside the quantum cases, because in every such case the 'symptomatic act' is in fact evidentially irrelevant to the event of which I'd taken it to be merely a sign. The upshot is at best a harmonious reconciliation between evidentialism and causalism, and at worst a de facto victory for the causalist.

This chapter aims to rebut the argument. More specifically, section 8.1 states its key premise in more detail, and also discusses the *dualism* between actions and observations that it involves. Section 8.2 asks whether anything in the nature of deliberation itself forces dualism upon us. The answer is no. Section 8.3 describes a general decision-theoretic difficulty that arises for any dualistic approach and section 8.4 sketches an anti-dualistic alternative that constitutes a more aggressive, and in my view the proper, evidentialist attitude.

8.1 Dualism and the Ramsey Thesis

The key premise of the argument is that a rational deliberating agent must take his presently contemplated acts to be evidentially irrelevant to

---

[3] Write $F(x) =_{\text{def.}} Cr(Ch(E|C) \leq x)$ and $F_C(x) =_{\text{def.}} Cr(Ch(E|C) \leq x\,|\,C)$. The probability calculus implies that $Cr(E|C) = \int_0^1 Cr(E|C \wedge Ch(E|C) = x)\,dF_C(x)$. If C is evidentially irrelevant to the past, including the chance function just before the time of C, then by the Principal Principle (2.26)(i), $\forall x F_C(x) = F(x)$. By this and the conditional version of the Principal Principle (2.26)(ii), it follows that $Cr(E|C) = \int_0^1 x\,dF(x) =$ the agent's expectation of the conditional chance of E given C. So the agent takes C to be evidentially relevant to E just in case she expects it to stand in what is in fact the relation of *causal* relevance to E.

anything then in his past.[4] I'll call this claim the **Ramsey Thesis (RT)**, Ramsey's statement of it being as follows:

> ... any possible present volition of ours is (for us) irrelevant to any past event. To another (or to ourselves in the future) it can serve as a sign of the past, but to us now what we do affects only the probability of the future.[5]

Notice that Ramsey does not state that the act ('volition') is evidentially irrelevant to the past for *everyone*. He says that this is so from the deliberating agent's point of view, but not from that of a spectator, who may even be the agent's future self. That is a plausible restriction. To think that one shouldn't draw any conclusions, or modify one's credences, about the past on the basis of the present actions of *another*, or the past actions of oneself, would be to deny e.g. that you should infer what they saw from the testimony of eyewitnesses, or that you had been drinking from your memory of having hit someone last night. More generally, given an option O and some possible past event S, there might be a strong statistical correlation between O-type events and preceding S-type events of which any 'mere' *observer* would be sensible to take account.

RT therefore distinguishes two ways of 'facing' the world: as observers, and as agents. According to it, aspects of the evidential bearing of an option O that are visible to an observer might be invisible to a rational *agent*, simply by virtue of her *being* an agent rather than an observer of O. Call this view **Evidential Dualism (ED)**.

I should emphasize that both Evidential Dualism and the Ramsey Thesis are here being understood as doctrines about the *deliberating* agent; that is, about someone who is choosing what to do, not someone who is *in the midst* of the action itself. Neither of them has anything to do with the idea that *whilst* you are doing something you have a *way* of knowing about it, at least under the description under which it is intentional; that is, different from the way in which anyone knows about it who is not doing it, for instance by ordinary observation.[6] That may be true, just as it is true that whilst you are seeing something you know about it in a way that is different from the way in which anyone knows about it who is not seeing it. And the explanation, in the case of action, is presumably neurological:

---

[4] Here and elsewhere in this chapter I am leaving out of consideration the quantum cases discussed in Chapter 6, on one interpretation of which even a causally undetermined option might remain evidentially correlated with events that are not in its future light cone.

[5] Ramsey 1990 [1929]: 158.

[6] Anscombe 1957: sections 8, 28–32. Also and especially O'Shaughnessy 2008: Chapter 9.

perhaps, that when we act we are directly aware not of the movements themselves, but of some pre-motor process that is associated with them.[7]

But *that* doctrine is irrelevant to the Ramsey Thesis, and irrelevant to decision theory. It distinguishes an agent's knowledge and an observer's knowledge of *what* the agent is doing *whilst* she is doing it. But (i) the Ramsey Thesis, Evidential Dualism, and subjective decision theory itself all concern an *earlier* state, i.e. when one is *deliberating* and so doesn't yet know what one is about to do.[8] (ii) The Ramsey Thesis and Evidential Dualism draw a distinction not between ways of knowing what the agent is doing from the two viewpoints, but between the evidential bearings from those viewpoints of one's actions *on other things*. I have nothing to say about the difference between the ways in which agent and observer can know what the agent is currently doing.

It is also worth emphasizing the distinction between Evidential Dualism and Evidential Decision Theory itself. Evidential Dualism is a claim about what you should *think*. It says that a rational agent contemplating an act will not take it to stand in the same relations of evidential relevance that she would if viewing it as an observer. Evidential Decision Theory itself is a claim about what you should *do*, the familiar claim that you should maximize *V*, and in itself makes no claims about the evidential connections that you can or should respect.

ED and EDT are therefore logically independent: one could hold that either is false whilst the other is true, that both are false, and that both are true. The connection between them is that if ED is true then EDT and CDT are plausibly equivalent. Is it?

8.2 Arguments for the Ramsey Thesis

The philosophers who have accepted ED typically do so on the grounds that a deliberating agent must take his present act to be completely free of external prior influences that affect its chances of occurrence. He must

---

[7] Haggard 2003: 119–24.

[8] Right after the quoted passage about 'any possible present volition of ours', Ramsey (1990 [1929]: 158) writes: 'This seems to me the root of the matter; that I cannot affect the past is a way of saying something quite clearly true about my degrees of belief. Again from the situation when we are deliberating seems to me to arise the general difference of cause and effect.' 'Again' suggests that 'the situation when we are deliberating' is what he had been discussing all along.

act on the assumption that not only determinism but also any mechanistic view of action is at least locally false.[9]

Thus Menzies and Price write that if O is an option then the conditional probability function $Cr(x|O)$, which reflects O's evidential bearing, is 'assessed from the agent's perspective under the supposition that the antecedent condition is realized *ab initio*, as a free act of the agent concerned'.[10] This is also the point of Ramsey's famous statement that 'my present action is an ultimate and the only ultimate contingency'.[11] The same idea is present in the work of Kant[12] and other philosophers.[13]

Note that what is supporting RT here cannot be indeterminism itself, at least not if we take RT to involve Evidential Dualism. If indeterminism is *simply* true in such a way that your acts, not being predictable, have no evidential bearing on anything other than what is in their future, then this is true for everyone, actor *or* observer. In that case there is no dualism. What is needed to make dualism plausible is not that indeterminism is true, but that it is in some way a *presupposition of the agent's perspective*. Thinking of your acts *as* causally undetermined is necessary for deliberating, even though you may know, from a more theoretical perspective, that they are in fact causally determined.

That thesis looks implausible. I myself think that determinism is both true and compatible with my acting freely. But I believe this all the time: I don't stop believing it every time I think about what next to say, write or do. None of the arguments for incompatibilism start to look convincing

---

9   Mechanism, which is weaker than strict determinism, says that 'human actions can be explained as the result of natural processes alone; that the mechanistic style of explanation, which works so well for electrons, motors, and galaxies, also works for us' (Bok 1998: 3).

10 Menzies and Price 1993: 190. Note that Menzies and Price take credences from the agent's perspective to be definitive of causal dependence and not answerable to some pre-existing conception of it. For concerns about this theory see section 5.3.1 above.

11 Ramsey 1990 [1929]: 158; for this interpretation of it see Price 1993: 261.

12 See Kant 2000 [1781/7]: A542/B570–A557/B585 on the 'transcendental idea of freedom', the point of which is to constitute 'a model of deliberative rationality which includes, as an ineliminable component, the thought of practical spontaneity . . . [T]he basic idea is simply that it is a condition of the possibility of taking oneself as a rational agent, that is, a being for whom reason is practical, that one attribute such spontaneity to oneself' (Allison 1990: 45). See also Kant 1964 [1785]: 124–6 (Ak. 4: 456–8) on the 'two standpoints'.

13 E.g. Hitchcock (1996: 519–22); Ismael (2007: section 4), and bizarrely also Pearl (2000: 108), who distinguishes the perspectives of observer and agent by calling the object of the first an 'act' and that of the second an 'action'. He then writes: 'An act is viewed from the outside, an action from the inside. Therefore, an act can be predicted and serve as evidence for the actor's stimuli and motivations (provided the actor is part of our model). Actions, in contrast, can neither be predicted nor provide evidence since (by definition) they are pending deliberation and turn into acts once executed.' I call this bizarre because Pearl scornfully rejects EDT as paradoxical: see Chapter 4 n. 5 above. But applied to what he calls 'actions', which are clearly its proper sphere, EDT is on this view of them bound to make the *same* recommendations as CDT.

when I start doing things, and then stop being convincing on reversion to my customary torpor.

One possible argument for RT, and so also for ED, is the well-known 'bilking' argument.[14] That argument runs as follows. It is generally true that one can, at least in principle, find out about the occurrence or otherwise of some past event independently of one's present intentions. So suppose that somebody claims that there is a correlation between events of a certain type S and *subsequent* acts of a type O. Then if O-type acts are in my power, I can always find out whether or not the S-type event has occurred, and then choose to perform the corresponding O-type act if and only if the S-type event has *not* occurred, thereby undermining or 'bilking' the supposed correlation between S and O, and hence also any evidential bearing that I took O to have upon S and which my conditional credences might otherwise have reflected. So I should not think, if the O-type event is in my present power, that it carries any news about the S-type past state. That is the bilking argument.

But when I learn that S is true, what I also learn is that I *learn* that S is true, and so my following the bilking policy need *not* undermine the correlation between S and O in cases where I am in fact ignorant of S. It therefore doesn't exclude my having, in the latter sort of case, a credential pattern on which $Cr(S|O) > Cr(S)$, but only one on which $Cr(S|O \wedge Cr(S) = 1) > Cr(S)$. For instance, in the standard Newcomb problem, my learning what the predictor has predicted will in *either* case motivate me to take both boxes, thus destroying any correlation between his prediction and what I do *in cases where I learn of the prediction*. But this is consistent with my thinking, in cases where I am ignorant of the prediction, that what I now do is evidentially relevant to what he has already predicted.[15]

Although the decision-theoretic literature contains little else by way of argument for RT,[16] recent work in the philosophy of mind suggests

---

[14] Flew 1954; Black 1956; Dummett 1964.

[15] This is not to deny what may be implausible for other reasons: that the bilking argument works against anyone who thinks that a presently contemplated option O might *cause* some past event S, at least if the obtaining of S is ascertainable independently of his present intention. For in that case $Cr(S|O) > Cr(S)$ entails $Cr(S|O \wedge Cr(S) = 1) > Cr(S)$, given that O screens off the agent's present beliefs (which are causally relevant to O) from the effects of O, including S. Dummett himself (1987: 294) is clear that the argument is at best only effective against backwards *causation*, and not against the evidential relevance of an option to *its own* earlier causes or to the side effects of those causes. But many cases over which EDT and CDT disagree fall into this latter category.

[16] Price (1991, 1993) in effect defends RT by taking it to be analytic of causation that an agent never takes her action to be evidence of anything except what causally depends on it. RT follows on the assumption that the past is causally independent of the present. I have already mentioned my concerns with this approach: see section 5.3.1.

another and more interesting line of thought. It exploits the *convertibility* of evidential bearing: if A has positive evidential bearing on B then B has positive evidential bearing on A. This is plausible in itself, and is anyway a consequence of the Bayesian conception of evidential bearing that I am taking for granted.[17] Hence if what I am about to do carries news (for me) of the past, then the past must itself carry news (for me) of what I am about to do. But nothing that I learn about the past *can* be evidence of what I am about to do, if I am rational: call this the **Converse Ramsey Thesis** (CRT). RT itself follows directly.

But is it true that no rational agent could regard any past state of affairs as evidence of what he is about to do, even if others do so regard it? One reason for believing so is the thesis that 'deliberation crowds out prediction': that whilst *deciding* what you are going to do, you cannot form *any* definite degree of belief about what you will in fact do.

The idea that deliberation and prediction are incompatible is not new,[18] and most of the concerns that it covers have been addressed elsewhere.[19] Here I will briefly discuss one very recent argument for their incompatibility. First, though, I should point out that the idea is really an objection to the whole approach to credence that I outlined at sections 1.4 and 2.3. That approach, as is mathematically entirely standard, defines *conditional* subjective probabilities of events upon acts $Cr(S|O)$ in terms that demand our making sense of the agent's unconditional credences in present options, since $Cr(S|O) =_{\text{def.}} Cr(S \wedge O)/Cr(O)$. And these probabilities figure essentially in the definition of Evidential Decision Theory itself. It *is* possible to explain conditional probabilities in some other way,[20] but in any case if the point stands then its relevance would extend well beyond that of the Converse Ramsey Thesis.

The most recent argument that deliberation crowds out prediction is due to Price, who claims that 'as she deliberates, an agent simply *does not have* knowledge, beliefs, or credences about the action in question'. 'One route to this thesis [he continues] turns on the special epistemic authority of deliberating agents concerning their own actions. This authority "trumps" any merely predictive knowledge claim about the same matters, making it necessarily unjustified.'[21]

---

[17] For proof see Chapter 2 n. 11.

[18] The 'crowding-out' formulation is due to Levi (1986: 66). For further discussion see Levi 1989; Gilboa 1999; Joyce 2002; Rabinowicz 2002; Ahmed 2007; Price 2012: 528–31.

[19] See especially Joyce 2002; Rabinowicz 2002.

[20] See e.g. Price 1986b; Edgington 1995: 266–7. Hájek (2003) takes conditional probabilities as primitive, and explains other probabilities in terms of them.

[21] Price 2012: 529. My emphasis.

The source of the authority is supposed to be the fact that a deliberator's beliefs about her own choices are generally evidence of their own truth.[22] If I am reasonably confident that I am going to smoke now, then presumably both that belief and my imminent smoking are effects of my now inclining in that direction, which is why all three tend to go together.

But just what authority *does* this give deliberators? If we are considering someone who has not yet firmly made up her mind, then her having a reasonably high confidence that she will do such-and-such hardly guarantees that she *will* do it, for since she has not yet made up her mind, the possibility must remain that she will change it. So the deliberator's intra-deliberative credences about her actions, if credences are what they are, cannot be *infallible*: that cannot be what her 'authority' consists in.

As far as I can see, the deliberator's authority consists in two things. First: the fact that her confidence that she will smoke is *generally correlated* with her being right about this. But then I can see no significant difference between this belief and beliefs on other matters on which she is generally reliable, for instance a high level of confidence that there is something red in her field of vision, or that yesterday was Thursday. But *these* fallible credences are bona fide credences if there are any at all, so why are not her similarly fallible, intra-deliberative credences about what she is going to do?

The second sense in which the deliberator's beliefs about her future actions are 'authoritative' is not that they are probably right, let alone infallible, but that we do not blame them for being wrong when they are wrong. This is the contrast, to which Anscombe brought modern philosophers' attention, between an error in the prediction and a shortfall in the performance. For instance, the difference between 'I am going to be sick' and 'I am going to take a walk' is that if in the first case I am not sick then I was wrong to think that I would be; but if in the second case I do not take a walk then my error lay not in thinking that I would take the walk, but in failing to do so.[23] But again, this sort of doxastic authority – or, better, exculpation – is no reason at all to think that what it calls authoritative are not genuine beliefs or credences that might be true or false, or more or less accurate.

It is in any case quite natural to think that the deliberating agent has credences about what she will do. Whilst deliberating about one matter

---

[22] Joyce 2007: 558; cited at Price 2012: 528. Joyce does not himself think that this precludes credences about one's own future acts.

[23] 'In some cases the facts are, so to speak, impugned for not being in accordance with the words, rather than *vice versa*. This is sometimes so when I change my mind...' – Anscombe 1957: section 2.

I might be asked to decide on another. And this might turn on my credences about the upshot of the first deliberative exercise, which may be going on concurrently (why not?). Whilst wondering whether to take a holiday in the last week of August, which I can book any time in the next six weeks, I get a chance, which will expire in five minutes, to purchase heavily discounted travel insurance for holidays taken in late August; but taking the insurance doesn't commit me to taking the holiday. Whether I purchase the insurance will and should depend on whether I think I will eventually decide to take the holiday. That is to say, it depends on my *credence* concerning a matter that is within my present field of deliberation. Certainly it is hard to see what else we should call the cognitive attitude towards my possible holiday that enters into my deliberations about the insurance. As I have just argued, whatever special authority that attitude has is no reason *not* to call it that.

So set aside the claim that deliberation crowds out prediction. Still, a line of thought that is related to Price's 'authority' argument might seem to supply more direct justification for the Converse Ramsey Thesis. This is the fact that persons who do treat such past states of affairs as evidence of what they are going to do will seem to have 'given up' on their own agency over those matters, and this is reason to think that rational control over one's contemplated acts is incompatible with admitting that sort of evidence for them.

Thus for an apparent instance of the Converse Ramsey Thesis, consider again the longtime gambler of section 3.2, the one who has made a wholly sincere resolution to stop.[24] He is walking past a casino: will he go in or not? Now imagine information E that he has made such resolutions many times before, and every time he has relapsed. This might be something that we spectators now learn. It might also be something that he had forgotten but now recollects. The new evidence E will certainly raise our confidence that he will in fact enter the casino.

But what are we to say if this gambler thinks to *himself, whilst* deliberating about whether to go back in: 'Oh well, I don't have any illusions: I've relapsed many times before; so probably, I will relapse again'? For him sincerely then to think that would be an evasion of his status as agent. For him, as for us, the character traits that his past backsliding reveals now appear as 'something independent of him, like a machine he has set

---

[24] The example adapts a famous one from Sartre (1969: 32 ff.) but this discussion of it owes more to Moran 2001: 78–83, which bases a similar point, and various other points, on a slightly different version of this story.

in motion and which now should carry him along without any further contribution from him'.[25]

The example *illustrates* the CRT, but its moral is quite generally applicable. For what lies behind our reaction here is that the question (a) 'Will I do this?' is normally understood as being *transparent to* the (practical, not ethical) question (b) 'Should I do it?' in the same sense in which the question (a) 'Do I believe that *p*?' has been called transparent to the question (b) '*Should* I believe that *p*?' or equivalently to 'Is it *the case* that *p*?' Transparency, as discussed at section 4.5, consists in the fact that I answer the (a)-question *by* answering the (b)-question. Thus if somebody asks me 'Do you believe that there will be a third world war?' I answer not by looking at e.g. the behavioural evidence that that is what I *do* believe, but precisely by looking at the evidence that there *will be* a third world war. And if we ask the gambler 'Are you going to visit this casino?' he should not look at his past behaviour for clues to the answer. Instead he should be asking whether gambling is now the thing *to* do.

Transparency accounts for the relative freedom with which *my* judgements of what I am going to do can respond to evidence that might for anyone else force the judgement one way or another. The reason the gambler's past backsliding should have no bearing on *his* question 'Am I going to enter the casino?' is that it has no bearing on the (b)-question to which it is transparent: 'Is that *the thing to do*?' More generally, it is rationally permissible for me *not* to respond to the news that *S* with the judgement that I will realize *O*, so long as *S* gives me no incentive to do so: whereas for another person, learning that *S* might well give *him* strong reason to think that I will in fact realize *O*.[26]

This line of thought looks like delivering the Converse Ramsey Thesis and hence also RT itself. At least, it makes it rationally *permissible* for an

---

[25] Moran 2001: 80. There, the 'machinery' that Moran intends is the gambler's past resolution, and not as I have it his dissolute character. But the point is the same.

[26] In cases where the past proposition *S does* incentivize *O*, the agent is *not* tempted to ignore it in his judgement as to whether he will realize *O*. For instance, in the standard Newcomb problem, your learning (S_1) that there is 1 million dollars in the opaque box will strongly incline you to think (O_2) that you will take both boxes because it is a decisive incentive to do just that. So this argument for the Converse Ramsey Thesis only applies to past events that do not incentivize or disincentivize *O*. But those other cases, where *S does* incentivize *O*, do not refute the Converse Ramsey Thesis, since your willingness to realize *O* upon learning *S* does not show that your $Cr(S|O) > Cr(O)$. For when you learn that *S* you also learn *that you learn* that *S*. So all it shows is that your $Cr(O|S \land Cr(S) = 1) > Cr(O)$. One might still hold that the CRT applies everywhere, including the Newcomb cases, but that we have this direct evidence for it only in cases where the past event does not constitute any incentive to realize one or another option.

agent *not* to treat as evidentially relevant to a contemplated act O some proposition S about the past that from an observer's perspective may be very strong evidence for O. This will be the case if S gives the agent no *incentive* to realize O. So it at least makes room for a contrast, between the evidential perspectives of agent and observer, of just the sort at which RT was aiming.

It looks that way. But both the Ramsey Thesis and its converse should create a feeling of unease that this argument for them does nothing to alleviate. How can it be, that when adopting the stance of an agent, we should just stop accepting the evidential relevance of statistical facts and theoretical calculations that remain what they always were? If accepting them is somehow an evasion of our status as agents, ignoring them is equally an evasion of our status as honest inquirers.

Betting on the Past (section 5.1) illustrates starkly how *ir*rational the Ramsey Thesis itself can seem. Before facing the problem, Alice knows perfectly well of all the evidence that speaks for a deterministic system of laws. Is she then supposed simply to *ignore* this evidence when asked to bet upon a proposition that determinism and that bet jointly entail? And if so, what could account for the intuition that she *should* bet that she was determined to bet in that way?

In any case, we can accommodate the undisputed facts about transparency without having to accept either the Ramsey Thesis or its converse. Notice first that the statement 'I am going to realize O', with O in my control, has two readings: (i) as an expression of my present intentions and (ii) as a genuinely future-tensed statement about what the upshot of my deliberations is in fact going to be. We need this distinction to explain why 'I am going to go for a walk – unless I decide not to' is not merely a tautology;[27] why 'I was going to go for a walk – but in the end I didn't' is not really a contradiction;[28] and perhaps also why, when I say 'I will go for a walk' and then do not go for a walk, the error is not in my earlier assertion but in my later performance. Normally the assertibility conditions for (i)- and (ii)-type statements coincide, because normally the deliberator *doesn't have* grounds for expecting that he will ultimately realize O other than his *present* intention to do it or not to do it; but in the extraordinary situations in which he does, they will diverge.

The sense of (a) 'Will I realize O?' that is transparent to the question (b) 'Should I?' is then the sense in which it is an invitation to give a (i)-type answer; that is, to express my *present* intention or inclination to do so. So

---

[27] Anscombe 1957: section 52.   [28] Dummett 1959: 21–2.

when I am answering the question in *this* sense it is hardly surprising that the only relevant 'evidence' consists of my *reasons* to realize O.[29] And it is both unsurprising and consistent with rationality that my inclination to assert 'I will realize O' on its type-(i) reading conforms to the Converse Ramsey Thesis, at least with respect to any past evidence for O that gives me no reason to do it or not to do it.

But nothing follows about whether it is a requirement of rational agency that the same evidential restrictions govern my inclination to assert 'I will realize O', or to believe the proposition that it expresses, when I take the latter on its type-(ii) or *pure future-tense* reading. For all that has been said, it could still be rational for the agent to think that although she is currently inclined towards not-O, this new bit of evidence, although not giving her any reason to choose O, is still grounds for expecting her evolving deliberation ultimately to settle on it.[30]

Thus consider the gambler. On recollecting his past backsliding, he might say: '(i) No, I will not gamble – at least that is not my current intention. But (ii) I know from experience of the to-and-fro of deliberation that these intentions are likely to change, and so it is likely that in the end I *will* gamble.' We should *not* understand anyone who asserts (ii) in this context as having somehow surrendered his agency to 'a machine he has set in motion and which now should carry him along without any further

---

[29] In fact there is a further distinction between two ways of approaching *this* question, i.e. the question about my present intentions. I could perhaps (α) approach it 'theoretically', i.e. attempt to describe my current 'standing' intentions; or I could (β) approach it 'deliberatively', i.e. attempt to *form* an intention in the light of what I currently take to speak for and against O (Moran 2001: section 2.5). In both cases I am doing something different from giving a type-(ii), i.e. purely future-tense, answer. What I am asserting here is that it is the (β)-type or *deliberative* approach that makes 'Will I realize O?' transparent to the question 'Should I?'

[30] This point does not apply to the Newcomb cases where the event S decisively incentivizes some option O. But as we have already seen (n. 26 above in this chapter), the agent's response to S in those cases gave no support to the Converse Ramsey Thesis, or therefore to RT itself, in the first place. The point is rather directed at those non-incentivizing cases in which RT has the most initial plausibility. If RT does not apply there then we have no reason to think that it applies to the Newcomb case either. Nor does the point apply to cases in which any past event has its evidential bearing on the agent's future exclusively via her present intentions or inclinations. In these cases, of which *P-Newcomb* at section 4.5 is an example, the agent's knowledge of her own intentions etc. ensures that $Cr(O|S) = Cr(O)$ for option O and past event S. But we already knew this, and also that in these cases EDT and CDT agree in recommending the dominant option. The cases that matter are those in which S has, at least from the observer's perspective, an evidential bearing upon O even once we are *given* the agent's present intentions: for instance, those that involve the propositions P and $\neg P$ in *Betting on the Past* (section 5.1), or that exploit the powers of the predictor in the standard Newcomb case. But the point does apply to them. If e.g. a soft determinist learns that the past state of the world was such as to cause her to realize an option O in five minutes, then she should think that that is what she will ultimately choose to do, whether or not that is what she presently intends.

contribution from him'. He *is* the machine, and what he is foreseeing is the upshot of his own deliberative mechanism. Nor does thinking of oneself in this way undermine one's own agency in any other way. To foresee that you will in fact *choose* an option O, whether or not this is something that you *now* want, is analytically not to see, or to foresee, that O is not a matter of choice at all.[31]

To summarize the discussion: the best evidence for the CRT, and hence for RT itself, is that you are certainly not more inclined to say 'I will realize O' upon learning of some past event S that anyone else would regard as evidence of O but which does nothing to incentivize that option. But that only means that learning S has no impact upon your current intentions. It *doesn't* mean that you either will or should ignore S's message that the future evolution of your deliberations will in fact terminate on this or that decision. For all that we have seen it is irrational to ignore *that* evidence, even in the midst of deliberation. And conversely, it is equally irrational, when deliberating, to ignore the evidential bearing of that ultimate choice upon the past. And it is the evidential and causal bearing of that ultimate choice, and not of any deliberation on the way to it, with which EDT and CDT are basically concerned.

So there are no good arguments for either the Ramsey Thesis or the Evidential Dualism that is its consequence. So far then, we should resist as gratuitous any such epistemological distinction between the 'agent's perspective' and the 'observer's perspective'. But in fact the position for ED, and therefore also for RT, is worse than that. If these perspectives *are* distinct then persisting beings like you and me must inevitably and successively take *both* perspectives on the *same* event. This gives rise to the more practical difficulty that the next section discusses.

8.3 Dynamic inconsistency and Dutch books

The following case combines a sanitized version of *Psycho Button A* (section 3.1) with something analogous to stage 2 of the *Newcomb-Insurance* problem (section 7.4.3). It is just for this reason that I am calling it:

> *Psycho-Insurance*: The problem is in two stages. In stage 1, you get to choose whether or not to press this button. If (S_1) the (Newcombian) predictor predicted that (O_1) you press it, then he has already arranged that doing so will cause one dollar to be debited from your bank account. If (S_2) he

---

[31] Russell (1926 [1914]: 238–40) makes the same point about imaginary beings that have *perfect* knowledge of both their future decisions and their future deliberative processes.

Table 8.1 *Psycho-Insurance* stage 1: dollar pay-offs

	S_1: predicts O_1	S_2: predicts O_2
O_1: press button	−1	1
O_2: don't press button	0	0

Table 8.2 *Psycho-Insurance* stage 2: dollar pay-offs

	T_1: predictor right	T_2: predictor wrong
P_1: bet	0.5	−1.5
P_2: don't bet	0	0

predicted that (O_1) you do *not* press it, then he has arranged that pressing it will cause one dollar to be *credited* to your account. If you were merely observing this scenario rather than acting in it, your confidence in the predictor's correctness, based on a vast amount of statistical data, would be $n > 0.75$. In that case you would have, and all equally informed observers do have, $Cr_1 (S_1|O_1) = Cr_1 (S_2|O_2) = n > 0.75$, here as at section 7.4.3 writing Cr_1 for your stage 1 credences.

Stage 2, of which the predictor neither knows nor cares, takes place after stage 1 but before you have had a chance to learn what the prediction was. You are offered the chance to bet $1.50 at odds of 1:3 that the predictor made a *correct* prediction at stage 1. As soon as you have made a decision, all is revealed.

The pay-offs in this scenario are as follows. At stage 1 we have Table 8.1. Dollar pay-offs for stage 2 are in Table 8.2. The amalgamated pay-offs for both stages are in Table 8.3.

Table 8.3 is derived from Tables 8.1 and 8.2 in the obvious way. For instance, if you take the options O_1 and P_1 and the predictor has predicted O_1, then you lose one dollar at stage 1 but win fifty cents at stage 2, for a net loss of fifty cents. Similarly, if the predictor has predicted O_2 then you win one dollar at stage 1 but lose $1.50 at stage 2, again for a net loss of fifty cents; and so on.

Let us first ask how an agent at stage 1 should *plan* to approach this problem of sequential choice. Looking at Table 8.3, we see that two such plans ($O_1 \wedge P_1$) and ($O_2 \wedge P_1$) are strictly dominated: whatever the

Table 8.3 Net pay-offs in
Psycho-Insurance problem

	S_1: predicts O_1	S_2: predicts O_2
$O_1 \wedge P_1$	-0.5	-0.5
$O_1 \wedge P_2$	-1	1
$O_2 \wedge P_1$	-1.5	0.5
$O_2 \wedge P_2$	0	0

predictor has predicted, $O_2 \wedge P_2$ does better than $O_1 \wedge P_1$; and what-ever he has predicted, $O_1 \wedge P_2$ does better than $O_2 \wedge P_1$. This doesn't rule out those plans unless S_1 and S_2 are relevantly independent of the plan. But for any evidentialist who holds RT, and also for CDT, this independence claim *does* hold: for the Ramseyan evidentialist because S_1 and S_2 are *evidentially* independent of the plan from the agential perspective that you currently occupy; for CDT because they are *causally* independent of it. So *ex ante*, anyone who accepts *either* RT and EDT, *or* CDT, will plan to carry out either $O_1 \wedge P_2$ or $O_2 \wedge P_2$.

The trouble is that whichever of these you plan *ex ante*, you will not be able to carry out the plan as soon as you reach stage 2. For once she *finds herself* at stage 2, the Ramseyan, and anyone else, would take option P_1; that is, would *accept* the bet. After all, you know that the predictor's strike rate exceeds 75 per cent, and you are being offered a bet at odds of 1:3 that the predictor has got it right this time. And at this stage of the problem, stage 2, there is no difficulty for any Evidential Dualist, or for anyone else, in accepting this evidence. At stage 2, the agent adopts an *observer's* perspective towards her choice at stage 1 and so is quite right by anybody's lights to heed its evidential bearing on the earlier prediction.

More specifically, suppose e.g. that at stage 2 you recall O_1, i.e. having pressed the button at stage 1. Then you *now* stand to this proposition O_1 as an observer, not as an agent. And from that perspective RT allows that the evidential bearing of O_1 upon S_1 both exists and is relevant. In this case $Cr_2 (T_1) = Cr_2 (S_1|O_1) = n > 0.75$, so it makes sense to accept the bet at stage 2. Similar reasoning applies if at stage 2 you recall O_2; also, in fact, if at stage 2 you are either uncertain of or wrong about what you did at stage 1.

So in this sequential problem the Ramsey Thesis forces the agent to violate the plausible principle that if a plan of action is rational then it should *remain* rational whilst you are implementing it. More formally, let an *n-stage sequential choice situation* Σ_n be an ordered n-tuple of sets Ω_i of options, $1 \leq i \leq n$. For any such sequential choice situation Σ_n let an *ex ante plan over* Σ_n, Π_n, be an n-tuple $(\pi_1, \ldots \pi_n)$ such that $\pi_i \in \Omega_i$. Finally, let the *plan continuation over* Σ_n of Π_n at stage k, Π_n^k be the $(n - k + 1)$-tuple $(\pi_k, \ldots \pi_n)$, $1 \leq k \leq n$. Then we may state the principle as follows:[32]

(8.3)   **Dynamic Consistency**: If Σ_n is a sequential choice situation and Π_n is an acceptable *ex ante* plan over Σ_n for an agent facing Σ_n, then if $1 \leq k \leq n$, Π_n^k is an acceptable plan continuation over Σ_n (for that agent) given that she has already implemented $\pi_1, \ldots \pi_{k-1}$.

RT violates this principle because for any agent that adheres to it, either (O_1, P_2) or (O_2, P_2) is an acceptable *ex ante* plan over the two-stage sequential choice situation $(\{O_1, O_2\}, \{P_1, P_2\})$. But their common continuation over that situation at stage 2, P_2, is *not* an acceptable continuation of either plan, given that the agent has implemented the first element of the plan. More informally again, EDT would recommend to the Ramseyan, and CDT would recommend to her and also to everyone else, a plan that either theory then tells you that you should, and which any rational person then would, drop halfway through carrying it out. This is surely a problem for RT, and also as it happens for CDT itself.

How much of a problem it is is perhaps most clearly visible in the case that your initial $Cr(S_1) \leq 0.5$. In that case RT supports $O_1 \wedge P_2$ at stage 1. But by the time you reach stage 2, it recommends P_1 instead of P_2. The upshot is therefore $O_1 \wedge P_1$ with a guaranteed net loss of fifty cents, when you could, by refusing all bets (i.e. $O_2 \wedge P_2$), have guaranteed a net loss of zero. So the *Psycho-Insurance* problem is in effect a 'Dutch book' against any Ramseyan whose credences initially satisfy the quite undemanding constraint that $Cr(S_1) \leq 0.5$. That is, it is a pattern of bets that this person will accept but which carry a guaranteed loss.[33]

---

[32] This is a simplified, weakened version of Verbeek's attractive principle of dynamic consistency (Verbeek 2008: 92), which is itself, as he notes (ibid.: 113 n. 21), a consequence in Hammond's original (1988) framework of whatever sort of rationality is required for the form of revealed preference that the latter accepts.

[33] The present argument resembles a diachronic 'Dutch Book Argument' (DBA) in that it faces the Ramseyan, or the Causal Decision Theorist, with a sequence of 'bets' that guarantee a fifty-cent loss, on which basis it rejects RT and CDT. This raises the concern that it may be vulnerable to criticisms that seem to undermine DBAs for probabilism (the doctrine, assumed true in this book, that the measure of rational beliefs is a probability function): for instance, that of Ramsey (1990 [1926]: 78). One problem for DBAs is that they are too pragmatic. For instance, it seems that the DBA for probabilism doesn't reveal any *epistemic* incoherence in beliefs that are not probability measures but only a *pragmatic* one. For all that it says, there might be nothing wrong with *having*

One might think to get around this essentially myopic predicament by means of the sophisticated approach that I discussed at section 7.4.3. In particular, if we start by thinking about stage 2, you know *ex ante* that you are going to take option P_1 by the time you reach it. So you should choose your options at stage 1 accordingly. In effect you are then choosing between $O_1 \wedge P_1$ and $O_2 \wedge P_1$. This doesn't solve the problem but only changes, without making any more implausible, the credences on which the problem arises. Suppose we write V_R for the news value that this *sophisticated Ramseyan* gives the options at stage 1. Then we have:

(8.4) $\quad U_1 (O_1 \wedge P_1) = V_R(O_1 \wedge P_1) = -0.5$

(8.5) $\quad U_1 (O_2 \wedge P_1) = V_R(O_2 \wedge P_1) = -1.5 Cr_1 (S_1) + 0.5 Cr_1 (S_2)$
$\quad\quad = 0.5 - 2 Cr_1 (S_1)$

So both CDT and RT endorse $O_1 \wedge P_1$ if $Cr (S_1) \geq 0.5$. So any sophisticated Ramseyan, or any sophisticated causalist, whose credences meet *that* undemanding condition is vulnerable to a guaranteed loss in *Psycho-Insurance*. Sophistication doesn't evade the sure loss. It simply changes the set of conditions under which the sure loss arises without making them any less plausible.

Note that in either case the difficulty is different from, but at least as serious as, that raised by the 'Why Ain'cha Rich' argument against CDT. The situation is not just that the Ramseyan with $Cr (S_1) \geq 0.5$, or the sophisticated Ramseyan with $Cr (S_1) \geq 0.5$, will foreseeably lose money in the long run relative to some alternative. It is that either sort of Ramseyan

---

beliefs that are not probability measures (for instance, in being more confident in P than in $P \vee Q$ for some propositions P, Q) as long as you don't *act* on them (Christensen 2004: 110). Whatever its justice against Ramsey's DBA, this argument has no relevance in the present context, in which the target is not (just) a pattern of beliefs but a precept for action. It is enough that *Psycho-Insurance* reveals a 'merely' pragmatic difficulty for RT and Causal Decision Theory. For if it is unwise to *act* on CDT then CDT is false; and if RT is pragmatically disastrous then it has no bearing on the dispute between EDT and CDT. A second issue for DBAs is their presupposition that if an agent prefers each one of a sequence or package of bets to not betting at all, then she prefers the combination of them to not betting at all. There is no obvious reason to think that rational agents are like that. Just because you would accept (b_1) a one-dollar bet on some proposition P at odds of 1:2, and (b_2) a one-dollar bet on $\neg P$ at the same odds, it doesn't follow that you would accept *either* bet, *given* that you had already taken on the other (Maher 1993: 96). Similarly a diachronic bettor who sees that taking b_1 will inevitably lead to her taking b_2 tomorrow, will refuse to take the first step on this path to certain loss, even though she might have accepted b_1 had b_2 not been in the offing tomorrow (Maher 1992: 124–5). But again, the difficulty does not arise in the present context. For in *Psycho-Insurance*, you know that you *will* bet at stage 2 *whatever* you do at stage 1. So from the stage 1 perspective of a sophisticated Ramseyan or causalist for whom $Cr_1(S_1) > 0.5$, taking O_1 at stage 1 is not the first step down the primrose path to ruin. It is doing the best that you now can to mitigate your future folly. (If you are not sophisticated but myopic then you are not going to concern yourself with your future decisions in any case.)

is *certain* to lose money *every time* she faces *Psycho-Insurance*. Worse still, the Ramseyan knows that she could have avoided loss altogether, since the sequence $O_2 \wedge P_2$ has a guaranteed net return of zero.

The reason that the Ramseyan evidentialist gets into this trouble is the Evidential Dualism that RT implies. I mean, that is what lies behind (a) the violation of dynamic consistency and (b) the vulnerability to guaranteed loss. After all, it is not as though you get any new and relevant evidence between stage 1 and stage 2, if by 'evidence' we mean known facts, e.g. the statistical data that underwrite subjective conditional credences. The statistical strike-rate of the predictor is known throughout. And after stage 1 you *learn* nothing new about your own situation that makes these especially relevant or irrelevant to it. For instance, it is irrelevant that you learn, at the end of stage 1, what you have chosen at that stage. Your decision at stage 2 is, and should be, to take the bet, even if you have forgotten what you did at stage 1. Nor do your desires ever change. Throughout the sequence your only concern is your terminal wealth.

What causes the problem is then not any change in beliefs or desires, but rather that one and the same proposition O_1 describes an *option*, or disjunction of options, from the *ex ante* perspective of the planner but only an observable *event* from the subsequent perspective that she reaches halfway through implementing the plan. Because she is an Evidential Dualist, the Ramseyan is therefore forced to take inconsistent attitudes towards that proposition depending on whether she is *ex ante* or *in media res*.

At stage 1 she must ignore the (known) statistical correlation between pushing the button and its having been predicted that she would. Focusing on the unsophisticated Ramseyan for the sake of illustration, the effect is that she plans at stage 1 to take what is effectively a bet that the predictor was wrong at odds of 1:1 if $Cr(S_1) \leq 0.5$, or no bet at all if $Cr(S_1) > 0.5$. But at stage 2, when what was once a potential act is now simply a state of the past, she must now bear this correlation in mind. And this generates a different evaluation of the remaining steps of the plan. More specifically, the effect is that the dualist, like everyone else, will see the offered bet at stage 2 as being on very generous terms. After all, she is arbitrarily confident that the predictor was correct at stage 1, and her now betting on that proposition does not have any (causal *or* evidential) bearing on this. So of course she will now bet that the predictor was *correct* at the offered odds of 1:3. The overall effect is at worst the fifty cent loss that $O_1 \wedge P_1$ guarantees, and at best the dynamic inconsistency that $O_2 \wedge P_1$ involves.

And the reason that the *causalist* gets into trouble is that CDT emulates Ramseyan dualism. In effect, it recommends acting *as if* you are an evidentialist to whom certain evidences become invisible just when, and only because, you are deliberating. In fact from the causalist perspective, whatever your initial credences and approach to sequential choice, *both* possible outcomes look very bad.

For instance, if you are sophisticated and have $Cr(S_1) < 0.5$, then you will end up with $O_2 \wedge P_1$. But as we see from Table 8.3, this sequence is dominated (by $O_2 \wedge P_1$) relative to a partition $\{S_2, S_2\}$ that is causally independent of anything that you now do: so you are throwing away fifty cents. Similarly, if you start with $Cr(S_1) > 0.5$ then you will end up with $O_1 \wedge P_1$, which is also thus dominated (by $O_2 \wedge P_2$). Finally, if your $Cr(S_1) = 0.5$ then CDT endorses both of these dominated sequences.[34]

This example is therefore also a final counterexample to CDT, regardless of what you think about the Ramsey Thesis. Whether or not she accepts RT, CDT leads the agent facing *Psycho-Insurance* into dynamic inconsistency and, if her credences meet a very light constraint, into certain loss.

An *anti*-dualistic EDT, which takes a consistent attitude towards the evidence at both stages, faces no such problems. At stage 1, $O_2 \wedge P_1$ will look best to any agent who follows EDT, whatever value her $Cr(S_1)$ then takes. And it will continue to look best at stage 2, when her $Cr(S_2)$ rises to n. So she will follow through on this plan for an expected return of $2n - 1.5$. No dynamic inconsistency, no guaranteed loss.

But before saying more about that alternative, let me informally sum up the negative message of this section. RT says 'that any possible *present* volition of ours is (for us) irrelevant to any past event. To another (*or to ourselves in the future*), it can serve as a sign of the past, but to us now what we do affects only the probability of the future'. That contrast between the two perspectives, in particular between the perspective of me now and the perspective of me in the future, is what forces me to take inconsistent attitudes towards the same proposition; and this is both characteristic of dualism and the source of these difficulties for it. Given also its lack of motivation, I think we should give up on the thought that the deliberating agent as such has some special kind of access to, or perspective upon, reality.

---

[34] One might question whether e.g. a sophisticated follower of CDT for whom $Cr_1(S_1) \geq 0.5$ will end up choosing O_1 at stage 1, on the grounds that this violates Joyce's principle of full information (3.7): when he learns that he is inclined towards O_1, this should increase that person's confidence that pressing the button will cause him to lose money and so will incline him back towards O_2. But this is not so: if $Cr_1(S_1) \geq 0.5$ then as you become more confident that you will take $O_1 \wedge P_1$, your $Cr_1(S_1)$ will increase – since you are confident of having been correctly predicted – and so by (8.5) your $U_1(O_2 \wedge P_1)$ will fall, thus reinforcing your confidence that you should take $O_1 \wedge P_1$. Similarly, if $Cr_1(S_1) \leq 0.5$, your increasing confidence that you will take $O_2 \wedge P_1$ will reduce your $Cr_1(S_1)$ and so your $U_1(O_2 \wedge P_1)$ will *rise*. So just as in the standard Newcomb case, what CDT recommends in *Psycho-Insurance* is stable under the reflection that that is what you are going to do.

It is both metaphysically extravagant and practically counterproductive. I now pass, briefly and finally, to its alternative.

8.4 Anti-dualism

To reject Evidential Dualism is not to deny the everyday facts from which it takes off. It is true that in real cases one almost never takes a contemplated act as symptomatic of the past. In cases where another person might do so, the reasons for the discrepancy are usually as stated at sections 4.1–4.5, i.e. because the agent already has background information that screens off the act from the past. But this has no essential connection with the agent's status *as* an agent. The difference is only that at least for the most part, the agent happens to possess this information and observers do not. But an observer might possess it too. If another person knew, for instance, that you had the same evidence as he did for thinking that you had overslept, then your hurrying to work would not be for him any further confirmation that you had overslept, any more than it is for you yourself.

What anti-dualism does deny is the quite unrestricted generalization that *no* presently contemplated act could ever be evidence of *any* past event, e.g. any past cause of it. The exceptions are cases where the *upshot* of deliberation is evidentially relevant to the past event quite independently of the agent's present evidence of her desires or her present inclination. Exceptions of this sort include *Betting on the Past* and the standard Newcomb problem, which for all its fantasy isn't actually incoherent. More realistic exceptions *may* include the version of *Prisoners' Dilemma* that I discussed at section 4.6. As I argued there, if the mechanism underlying consensus effect is that one takes the *entire course* of one's deliberation to simulate, with more or less accuracy, the entire course of others' deliberations, then Alice will take the upshot of her deliberation to be evidentially relevant to the upshot of Bob's even given her knowledge of her present inclinations. In application to these cases, the dualistic generalization is unmotivated and creates both dynamic inconsistency and certain loss.

According to anti-dualism the deliberating agent occupies *no* special perspective: in particular, none from which the acts that he is contemplating cease to bear upon past events with which he knows them to be correlated. The evidential bearing of a future act on a past event is the same, whether or not it is under your control, if and when the empirical evidence for its bearing thus on the past is the same. In so far as we as agents ignore whatever evidence we respect as observers, this is not something that our agential perspective vindicates. It is just a mistake.

Anti-dualism is thus a Humean doctrine,[35] and can seem open to the charge that philosophers have laid against Hume himself, of presenting the human subject as a merely passive observer of the passing show.[36] But whatever its justice against Hume, that is an overreaction to anti-dualism. Nobody is denying that when their bodies move in certain ways people really are *doing* things, that they are doing them because they want to do them, or even that the people who plan to do them are usually then in a better position to predict them than is anyone else. Anti-dualism only denies that any agent must *ipso facto and always* think that she is epistemically unique with regard to her contemplated future acts. On the contrary, it allows that her future acts *may* have evidential bearing on past or concurrent events to which observers might also have independent access. This is not to reject anything about agency worth wanting, but only to resist a kind of metaphysical inflation of it.

As a final commendation of it let me point out that anti-dualism also shares with Hume's general outlook an obviously naturalistic motivation. By 'naturalism' I mean the regulative ideal that takes us to be entirely of a piece with nature. More specifically, it is the aspiration to explain the human world and everything in it in terms of principles, laws and models that also apply elsewhere and ideally as broadly as possible. The religious and metaphysical systems of the past typically, and many now-discredited scientific theories in fact, all more or less consciously rejected naturalism so understood. In their own ways biblical Creationism, geocentrism, Cartesian dualism, Hegelianism and vitalism all gave some privileged or exceptional place to humanity or to some feature of its life. Evidential Dualism rejects it too: in attempting to accommodate the prescriptions of CDT it recommends that an agent treat her own contemplated acts as

---

[35] In connection with the unpredictability of action he writes that 'though in reflecting on human actions we seldom feel such a looseness or indifference, yet it very commonly happens, that in performing the actions themselves we are sensible of something like it . . . We *feel* that our actions are subject to our will on most occasions, and *imagine we feel* that the will itself is subject to nothing; because when by a denial of it we are provoked to try, we feel that it moves easily every way, and produces an image of itself even on that side, on which it did not settle . . . But these efforts are all in vain; and whatever capricious and irregular actions we may perform; as the desire of showing our liberty is the sole motive of our actions; we can never free ourselves from the bonds of necessity. We may imagine we feel a liberty within ourselves; but a spectator can commonly infer our actions from our motives and character; and even where he cannot, he concludes in general, that he might, were he perfectly acquainted with every circumstance of our situation and temper, and the most secret springs of our complexion and disposition.' Hume 1949 [1738]: II.iii.2. My emphasis. Here he appeals to the spectator's vision as a *corrective* to the agent's, thus rejecting the idea that when I act I see my actions from a different *perspective* than does a spectator.

[36] Flew 1954: 49–50; Ritchie 1967: 90 ff.; Menzies and Price 1993: 191.

being for some reason exempt from empirical standards that apply everywhere else. By contrast, the anti-dualistic Evidential Decision Theory that I am recommending in its place must inevitably clash with CDT over cases that are certainly coherent and may not be marginal. This more aggressive stance towards Causal Decision Theory is the inevitable cost – or rather, and as I have tried to show, benefit – of seeing your own actions in the same scientific light that you should also turn upon everything else.

References

Ahmed, A. 2005. Evidential decision theory and medical Newcomb problems. *British Journal for the Philosophy of Science* 56: 191–8.

2007. Agency and causation. In R. Corry and H. Price (eds.), *Causation, Physics and the Constitution of Reality: Russell's Republic Revisited*. Oxford: Oxford University Press: 120–55.

2013. Causal Decision Theory: a counterexample. *Philosophical Review* 122: 289–306.

Allais, M. 1953. Le comportement de l'homme rationnel devant le risque: Critique des postulats et axiomes de l'école Américaine. *Econometrica* 21: 503–46.

Allison, H. E. 1990. *Kant's Theory of Freedom*. Cambridge: Cambridge University Press.

Anand, P. 1990. Two types of utility. *Greek Economic Review* 12: 58–74.

Anscombe, G. E. M. 1957. *Intention*. Oxford: Blackwell.

Arnett, J. J. 2000. Optimistic bias in adolescent and adult smokers and nonsmokers. *Addictive Behaviours* 25: 625–32.

Arntzenius, F. 2008. No regrets, or: Edith Piaf revamps decision theory. *Erkenntnis* 68: 277–97.

Arntzenius, F., A. Elga and J. Hawthorne. 2004. Bayesianism, infinite decisions and binding. *Mind* 113: 251–83.

Ayer, A. J. 1963. Fatalism. In his *Concept of a Person and Other Essays*. London: Macmillan: 235–68.

Bach, K. 1987. Newcomb's problem: the $1,000,000 solution. *Canadian Journal of Philosophy* 17: 409–25.

Beebee, H. 2003. Local miracle compatibilism. *Noûs* 37: 258–77.

Beebee, H. and A. Mele. 2002. Humean compatibilism. *Mind* 442: 201–23.

Beebee, H. and D. Papineau. 1997. Probability as a guide to life. *Journal of Philosophy* 94: 217–43. Reprinted in D. Papineau, *The Roots of Reason*. Oxford: Oxford University Press, 2003: 130–66.

Bell, J. S. 1964. On the Einstein–Podolsky–Rosen paradox. *Physics* 1: 195–200. Reprinted in his *Speakable and Unspeakable in Quantum Mechanics*. Revised edition. Cambridge: Cambridge University Press, 2004: 14–21.

1981. Bertlmann's socks and the nature of reality. *Journal de physique* 42: C2 41–61. Reprinted in *Speakable and Unspeakable in Quantum Mechanics*. Revised edition. Cambridge: Cambridge University Press, 2004: 139–58.

1985. The theory of local beables. *Dialectica* 39: 85–96. Reprinted in *Speakable and Unspeakable in Quantum Mechanics*. Revised edition. Cambridge: Cambridge University Press, 2004: 52–62.

1990. La nouvelle cuisine. In A. Sarlemijn and P. Kroes (eds.), *Between Science and Technology*. Amsterdam: Elsevier: 97–115. Reprinted in *Speakable and Unspeakable in Quantum Mechanics*. Revised edition. Cambridge: Cambridge University Press, 2004: 323–48.

Bennett, J. 2003. *A Philosophical Guide to Conditionals*. Oxford: Oxford University Press.

Berkeley, G. 1980 [1710]. *Principles of Human Knowledge*. In his *Philosophical Works*, ed. M. Ayers. London: Everyman.

Berkovitz, J. 1995. Quantum nonlocality: an analysis of the implications of Bell's theorem and quantum correlations for nonlocality. Ph.D. thesis, University of Cambridge.

Bermudez, J. L. 2013. Prisoner's dilemma and Newcomb's problem: why Lewis's argument fails. *Analysis* 73: 423–9.

Binmore, K. 2009. *Rational Decisions*. Princeton: Princeton University Press.

Black, M. 1956. Why cannot an effect precede its cause? *Analysis* 16: 49–58.

Blaylock, G. 2010. The EPR paradox, Bell's inequality, and the question of locality. *American Journal of Physics* 78: 111–20.

Bok, H. 1998. *Freedom and Responsibility*. Princeton: Princeton University Press.

Bolker, E. 1967. A simultaneous axiomatization of utility and subjective probability. *Philosophy of Science* 34: 333–40.

Bovens, L. and S. Hartmann. 2003. *Bayesian Epistemology*. Oxford: Oxford University Press.

Broome, J. 1989. An economic Newcomb problem. *Analysis* 49: 220–2.

1992. Review of McClennen, *Rationality and Dynamic Choice: Foundational Explorations*. Cambridge: Cambridge University Press, 1990. *Ethics* 102: 666–8.

Butterfield, J. 1992a. Bell's theorem: what it takes. *British Journal for the Philosophy of Science* 43: 41–83.

1992b. David Lewis meets John Bell. *Philosophy of Science* 59: 26–42.

Cantwell, J. 2013. Conditionals in causal decision theory. *Synthese* 190: 661–79.

Cartwright, N. 1979. Causal laws and effective strategies. *Noûs* 13: 419–37.

1999. Causal diversity and the Markov condition. *Synthese* 121: 3–127.

Cavalcanti, E. 2010. Causation, decision theory and Bell's theorem: a quantum analogue of the Newcomb problem. *British Journal for the Philosophy of Science* 61: 569–97.

Christensen, D. 2004. *Putting Logic in Its Place*. Oxford: Oxford University Press.

Chu, F. and J. Y. Halpern. 2004. Great expectations. Part II: generalized expected utility as a universal decision rule. *Artificial Intelligence* 159: 207–29.

Clark, M. 2007. *Paradoxes from A to Z*. 2nd edition. London: Routledge.

Clauser, J. F., M. A. Horne, A. Shimony and R. A. Holt. 1969. Proposed experiment to test local hidden-variable theories. *Physical Review Letters* 23: 880–4.

Collingwood, R. G. 1940. *Essay on Metaphysics*. Oxford: Oxford University Press.

Craig, W. L. 1987. Divine foreknowledge and Newcomb's paradox. *Philosophia* 17: 331–50.

Davidson, D. 1971. Agency. In R. Brinkley, R. Bronaugh and A. Marras (eds.), *Agent, Action and Reason*. Toronto: University of Toronto Press: 3–25. Reprinted in his *Essays on Actions and Events*. 2nd edition. Oxford: Oxford University Press, 2001: 43–61.

Dawes, R. M. 1990. The potential nonfalsity of the false consensus effect. In H. J. Einhorn and R. M. Hogarth (eds.), *Insights in Decision Making: A Tribute to Hillel J. Einhorn*. Chicago: The University of Chicago Press: 179–99.

Dougherty, T. 2011. On whether to prefer pain to pass. *Ethics* 121: 521–37.

 2013. A deluxe money pump. *Thought*. Online, available at http://onlinelibrary. wiley.com/doi/10.1002/tht3.91/full.

Downs, A. 1957. *An Economic Theory of Democracy*. New York: Harper and Row.

Dummett, M. A. E. 1959. Truth. *Proceedings of the Aristotelian Society* 59: 141–62. Reprinted in his *Truth and Other Enigmas*. London: Duckworth, 1978: 1–24.

 1964. Bringing about the past. *Philosophical Review* 73: 338–59. Reprinted in *Truth and Other Enigmas*. London: Duckworth, 1978: 333–50.

 1986. Causal loops. In R. Flood and M. Lockwood (eds.), *The Nature of Time*. Oxford: Blackwell: 135–69. Reprinted in M. A. E. Dummett, *The Seas of Language*. Oxford: Oxford University Press, 1993: 349–75.

 1987. Reply to D. H. Mellor. In B. Taylor (ed.), *Michael Dummett: Contributions to Philosophy*. Dordrecht: Martinus Nijhoff: 287–98.

Edgington, D. 1995. On conditionals. *Mind* 104: 235–329.

 2004. Counterfactuals and the benefit of hindsight. In P. Dowe and P. Noordhof (eds.), *Cause and Chance*. London: Routledge: 12–27.

 2011. Conditionals, causation and decision. *Analytic Philosophy* 52: 75–87.

Edgley, R. 1969. *Reason in Theory and Practice*. London: Hutchinson.

Eells, E. 1982. *Rational Decision and Causality*. Cambridge: Cambridge University Press.

Egan, A. 2007. Some counterexamples to Causal Decision Theory. *Philosophical Review* 116: 93–114.

Elga, A. 2007. Reflection and disagreement. *Noûs* 41: 478–502.

Evans, G. 1982. *The Varieties of Reference*. Oxford: Oxford University Press.

Fischer, J. M. 1994. *The Metaphysics of Free Will: An Essay on Control*. Oxford: Oxford University Press.

Fishburn, P. 1970. *Utility Theory for Decision Making*. New York: John Wiley and Sons.

Flew, A. G. N. 1954. Can an effect precede its cause? *Proceedings of the Aristotelian Society Supplementary Volume* 28: 47–62.

Forster, M. N. 2010. Wittgenstein on family resemblance concepts. In A. Ahmed (ed.), *Wittgenstein's 'Philosophical Investigations': A Critical Guide*. Cambridge: Cambridge University Press: 66–87.

Frydman, R., G. P. O'Driscoll and A. Schotter. 1982. Rational expectations of government policy: an application of Newcomb's problem. *Southern Economic Journal* 1982: 311–19.

Gardner, M. 1974. Mathematical games. *Scientific American* 230: 102–9.

Gibbard, A. 1992. Weakly self-ratifying strategies: comments on McClennen. *Philosophical Studies* 65: 217–25.

Gibbard, A. and W. L. Harper. 1978. Counterfactuals and two kinds of expected utility. In C. Hooker, J. Leach and E. McClennen (eds.), *Foundations and Applications of Decision Theory*. Dordrecht: Riedel: 125–62. Reprinted in P. Gärdenfors and N.-E. Sahlin (eds.), *Decision, Probability and Utility*. Cambridge: Cambridge University Press, 1988: 341–76.

Gilboa, I. 1999. Can free choice be known? In C. Bicchieri, R. Jeffrey and B. Skyrms (eds.), *The Logic of Strategy*. Oxford: Oxford University Press: 163–74.

  2009. *Theory of Decision under Uncertainty*. Cambridge: Cambridge University Press.

Gilboa, I. and D. Schmeidler. 1989. Maxmin expected utility with a non-unique prior. *Journal of Mathematical Economics* 18: 141–53.

Glymour, C. 1980. *Theory and Evidence*. Princeton: Princeton University Press.

Grafstein, R. 1991. An evidential decision theory of turnout. *American Journal of Political Science* 35: 989–1010.

  1999. *Choice-Free Rationality: A Positive Theory of Political Behaviour*. Ann Arbor: University of Michigan Press.

Greaves, H. 2004. Understanding Deutsch's probability in a deterministic multiverse. *Studies in History and Philosophy of Modern Physics* 38: 120–52.

  2007. Probability in the Everett interpretation. *Philosophy Compass* 2: 109–28.

Green, D. P. and I. Shapiro 1994. *Pathologies of Rational Choice Theory*. New Haven: Yale University Press.

Haggard, P. 2003. Conscious awareness of intention and action. In J. Roessler and N. Eilan (eds.), *Agency and Self-Awareness*. Oxford: Oxford University Press: 111–27.

Hájek, A. 2003. What conditional probability could not be. *Synthese* 137: 273–323.

Hammond, P. 1988. Consequentialist foundations for expected utility theory. *Theory and Decision* 25: 25–78.

Hampshire, S. 1975. *Freedom of the Individual*. London: Chatto and Windus.

Hausman, D. 1999. Lessons from quantum mechanics. *Synthese* 121: 79–92.

Hedden, B. 2012. Options and the subjective *ought*. *Philosophical Studies* 158: 343–60.

Hitchcock, C. 1996. Causal decision theory and decision-theoretic causality. *Noûs* 30: 508–26.

  2004. Do all and only causes raise the probabilities of effects? In J. Collins, N. Hall and L. A. Paul (eds.), *Causation and Counterfactuals*. Cambridge, MA: MIT Press: 403–17.

  2007. What Russell got right. In R. Corry and H. Price (eds.), *Causation, Physics and the Constitution of Reality*. Oxford: Oxford University Press: 45–65.

Horgan, T. 1981. Counterfactuals and Newcomb's problem. *Journal of Philosophy* 78: 331–56.

1985. Newcomb's problem: a stalemate. In R. Campbell and L. Sowden (eds.), *Paradoxes of Rationality and Cooperation: Prisoner's Dilemma and Newcomb's Problem.* Vancouver: University of British Columbia Press: 223–34.

Horwich, P. 1987. *Asymmetries in Time: Problems in the Philosophy of Science.* Cambridge, MA: MIT Press.

Hubin, D. and G. Ross. 1985. Newcomb's perfect predictor. *Noûs* 19: 439–46.

Hume, D. 1949 [1738]. *Treatise of Human Nature.* London: J. M. Dent.

1975 [1777]. *Enquiries Concerning Human Understanding and Concerning the Principles of Morals.* Ed. L. A. Selby-Bigge. Oxford: Clarendon Press.

Ismael, J. 2007. Freedom, compulsion and causation. *Psyche* 13.

Jarrett, J. P. 1984. On the physical significance of the locality conditions in the Bell arguments. *Noûs* 18: 569–89.

Jaynes, E. T. 2003. *Probability Theory: The Logic of Science: Principles and Elementary Applications vol. 1.* Cambridge: Cambridge University Press.

Jeffrey, R. C. 1965. *The Logic of Decision.* 1st edition. Chicago: The University of Chicago Press.

1977. Savage's Omelet. In *PSA: Proceedings of the Biennial meeting of the Philosophy of Science Association 1976*: 361–71.

1983. *The Logic of Decision.* 2nd edition. Chicago: The University of Chicago Press.

Joyce, J. 1999. *Foundations of Causal Decision Theory.* Cambridge: Cambridge University Press.

2002. Levi on Causal Decision Theory and the possibility of predicting one's own actions. *Philosophical Studies* 110: 69–102.

2007. Are Newcomb problems really decisions? *Synthese* 156: 537–62.

2012. Regret and instability in Causal Decision Theory. *Synthese* 187: 123–45.

Kahneman, D. and A. Tversky. 1981. The framing of decisions and the psychology of choice. *Science* 1981: 453–8.

Kant, I. 1964 [1785]. *Groundwork of the Metaphysics of Morals.* Tr. H. J. Paton. New York: Harper and Row.

2000 [1781/7]. *Critique of Pure Reason.* Ed. and tr. P. Guyer and A. W. Wood. Cambridge: Cambridge University Press.

Koestner, R., G. F. Losier, N. M. Worren, L. Baker and R. J. Vallerand. 1995. False consensus effects for the 1992 Canadian referendum. *Canadian Journal of Behavioural Science* 27: 214–25.

Knox, R. E. and J. A. Inkster. 1998. Postdecision dissonance at post time. *Journal of Personality and Social Psychology* 8: 319–23.

Kraft, C. H., J. W. Pratt and A. Seidenberg. 1959. Intuitive probability on finite sets. *Annals of Mathematical Statistics* 30: 408–19.

Kreps, D. M. 1988. *Notes on the Theory of Choice.* Boulder, CO: Westview.

1990. *Game Theory and Economic Modelling.* Oxford: Oxford University Press.

Kripke, S. 1980. *Naming and Necessity.* Oxford: Blackwell.

Krueger, J. and J. S. Zeiger. 1993. Social categorization and the truly false consensus effect. *Journal of Personality and Social Psychology* 65: 670–80.

Kvart, I. 2004. Probabilistic cause, edge conditions, late preemption, and discrete cases. In P. Dowe and P. Noordhof (eds.), *Cause and Chance: Causation in an Indeterministic World*. London: Routledge: 163–87.

Kydland, F. E. and E. C. Prescott. 1977. Rules rather than discretion: the inconsistency of optimal plans. *Journal of Political Economy* 85: 473–91.

Laudisa, F. 2001. Non-locality and theories of causation. In J. Butterfield and T. Placek (eds.), *Non-locality and Modality*. Dordrecht: Kluwer: 223–34.

Ledwig, M. 2000. Newcomb's problem. Dissertation submitted to the University of Constance.

Leeds, S. 1984. Eells and Jeffrey on Newcomb's problem. *Philosophical Studies* 46: 97–107.

Leslie, J. 1991. Ensuring two bird deaths with one throw. *Mind* 100: 73–86.

Levi, I. 1974. On indeterminate probabilities. *Journal of Philosophy* 56: 391–418.

1975. Newcomb's many problems. *Theory and Decision* 6: 161–75.

1985. Common causes, smoking and lung cancer. In R. Campbell and L. Sowden (eds.), *Paradoxes of Rationality and Cooperation: Prisoner's Dilemma and Newcomb's Problem*. Vancouver: University of British Columbia Press: 234–47.

1986. *Hard Choices*. Cambridge: Cambridge University Press.

1989. Rationality, prediction and autonomous choice. *Canadian Journal of Philosophy*. Supp. Vol. 19: 339–63. Reprinted in his *Covenant of Reason: Rationality and the Commitments of Thought*. Cambridge: Cambridge University Press, 1997: 19–39.

1992. Consequentialism and sequential choice. In M. Bacharach and S. Hurley (eds.), *Foundations of Decision Theory*. Oxford: Blackwell: 92–122. Reprinted in I. Levi, *The Covenant of Reason*. Cambridge: Cambridge University Press, 1997: 70–101.

Lewis, D. K. 1973a. Causation. *Journal of Philosophy* 70: 556–67. Reprinted in his *Philosophical Papers vol. 2* Oxford: Oxford University Press, 1986: 159–213.

1973b. *Counterfactuals*. Oxford: Blackwell.

1979a. Counterfactual dependence and time's arrow. *Noûs* 13: 455–76. Reprinted in his *Philosophical Papers vol. 2*. Oxford: Oxford University Press, 1986: 32–66.

1979b. Prisoners' dilemma is a Newcomb problem. *Philosophy and Public Affairs* 8: 235–40. Reprinted in his *Philosophical Papers vol. 2*. Oxford: Oxford University Press, 1986: 299–304.

1980. A subjectivist's guide to objective chance. In R. Jeffrey, ed., *Studies in Inductive Logic and Probability, vol. 2*. Berkeley and Los Angeles: California University Press: 263–93. Reprinted in his *Philosophical Papers vol. 2*. Oxford: Oxford University Press, 1986: 83–132.

1981a. Are we free to break the laws? *Theoria* 47: 113–21. Reprinted in his *Philosophical Papers vol. 2*. Oxford: Oxford University Press, 1986: 291–8.

1981b. Causal decision theory. *Australasian Journal of Philosophy* 59: 5–30. Reprinted in his *Philosophical Papers vol. 2*. Oxford: Oxford University Press, 1986: 305–39.

1981c. Why ain'cha rich? *Noûs* 15: 277–80. Reprinted in his *Papers in Ethics and Social Philosophy*. Cambridge: Cambridge University Press, 2000: 37–41.

1983. Levi against U-maximization. *Journal of Philosophy* 80: 531–4.

1986. Events. In his *Philosophical Papers vol. 2*. Oxford: Oxford University Press: 241–69.

Locke, J. 1975 [1689]. *Essay Concerning Human Understanding*. Ed. P. H. Nidditch. Oxford: Oxford University Press.

Loewer, B. 2007. Counterfactuals and the second law. In R. Corry and H. Price (eds.), *Causation, Physics and the Constitution of Reality*. Oxford: Oxford University Press: 293–326.

McClennen, E. F. 1990. *Rationality and Dynamic Choice: Foundational Explorations*. Cambridge: Cambridge University Press.

McFetridge, I. 1990. Logical necessity: some issues. In his *Logical Necessity and Other Essays*. Ed. J. Haldane and R. Scruton. London: Aristotelian Society: 135–54.

McKay, P. K. 2007. Freedom, fiction and Evidential Decision Theory. *Erkenntnis* 66: 393–407.

Mackie, J. L. 1980. *The Cement of the Universe*. Oxford: Clarendon Press.

Maher, P. 1990. Symptomatic acts and the value of evidence in Causal Decision Theory. *Philosophy of Science* 57: 479–98.

1992. Diachronic rationality. *Philosophy of Science* 59: 120–41.

1993. *Betting on Theories*. Cambridge: Cambridge University Press.

Maudlin, T. 2002. *Quantum Non-locality and Relativity*. Oxford: Blackwell.

Meacham, C. 2010. Binding and its consequences. *Philosophical Studies* 149: 49–71.

Meek, C. and C. Glymour. 1994. Conditioning and intervening. *British Journal for the Philosophy of Science* 45: 1001–21.

Mellor, D. H. 1983. Objective decision making. *Social Theory and Practice* 9: 289–309. Reprinted in his *Matters of Metaphysics*. Cambridge: Cambridge University Press, 1991: 269–87.

1987. Fixed past, unfixed future. In B. Taylor (ed.), *Michael Dummett: Contributions to Philosophy*. Dordrecht: Martinus Nijhoff: 166–84.

1993. How to believe a conditional. *Journal of Philosophy* 60: 233–48.

2005. What does subjective decision theory tell us? In H. Lillehammer and D. H. Mellor (eds.), *Ramsey's Legacy*. Oxford: Oxford University Press: 137–48.

Menzies, P. and H. Price. 1993. Causation as a secondary quality. *British Journal for the Philosophy of Science* 44: 187–203.

Mermin, N. 1981. Quantum mysteries for anyone. *Journal of Philosophy* 78: 397–408.

Michotte, A. 1963. *The Perception of Causality*. London: Methuen.

Mill, J. S. 1952 [1843]. *A System of Logic*. 8th edition. London: Longmans, Green and Co.

Moran, R. 2001. *Authority and Estrangement*. Princeton: Princeton University Press.

Morton, A. 2002. *The Importance of Being Understood: Folk Psychology as Ethics*. London: Routledge.

Muckenheim, W. 1982. A resolution of the Einstein–Podolsky–Rosen paradox. *Lettere al Nuovo Cimento* 35: 300–4.

Mullen, B., J. L. Atkins, D. S. Champion, C. Edwards, D. Hardy and J. E. Story. 1985. The false consensus effect: a meta-analysis of 115 hypothesis tests. *Journal of Experimental Social Psychology* 21: 262–83.

Muth, J. F. 1961. Rational expectations and the theory of price movements. *Econometrica* 29: 315–35.

Nagel, T. 1970. *The Possibility of Altruism*. Oxford: Oxford University Press.

Nozick, R. 1969. Newcomb's problem and two principles of choice. In N. Rescher (ed.), *Essays in Honor of Carl G. Hempel*. Dordrecht: D. Reidel: 114–46. Reprinted in P. Moser (ed.), *Rationality in Action: Contemporary Approaches*. Cambridge: Cambridge University Press, 1990: 207–34.

　1993. *The Nature of Rationality*. Princeton: Princeton University Press.

Osborne, M. J. and A. Rubinstein. 1994. *A Course in Game Theory*. Cambridge, MA: MIT Press.

O'Shaughnessy, B. 2008. *The Will: A Dual Aspect Theory*. 2nd editon. 2 vols. Cambridge: Cambridge University Press.

Parfit, D. 1984. *Reasons and Persons*. Oxford: Oxford University Press.

Pearl, J. 2000. *Causality: Models, Reasoning and Inference*. Cambridge: Cambridge University Press.

Peterson, M. 2009. *An Introduction to Decision Theory*. Cambridge: Cambridge University Press.

Pollock, J. L. 2010. A resource-bounded agent addresses the Newcomb problem. *Synthese* 176: 57–82.

Price, H. 1986a. Against Causal Decision Theory. *Synthese* 67: 195–212.

　1986b. Conditional credence. *Mind* 95: 18–36.

　1991. Agency and probabilistic causality. *British Journal for the Philosophy of Science* 42: 157–76.

　1993. The direction of causation: Ramsey's ultimate contingency. *Philosophy of Science Association* 2: 253–67.

　2010. Decisions, decisions, decisions: can Savage salvage Everettian probability? In S. Saunders, J. Barrett, A. Kent and D. Wallace (eds.), *Many Worlds? Everett, Quantum Theory and Reality*. Oxford: Oxford University Press: 369–90.

　2012. Causation, chance, and the rational significance of supernatural evidence. *Philosophical Review* 121: 483–538.

Quattrone, G. A. and A. Tversky. 1986. Self-deception and the voter's illusion. In J. Elster (ed.), *The Multiple Self*. Cambridge: Cambridge University Press: 35–58.

Quine, W. V. 1981. What price bivalence? *Journal of Philosophy* 78: 90–5. Reprinted in his *Theories and Things*. Cambridge, MA: Harvard University Press 1981: 31–7.

Rabinowicz, W. 2000. Money pump with foresight. In M. J. Almeida (ed.), *Imperceptible Harms and Benefits*. Dordrecht: Kluwer: 123–54.

2002. Does practical deliberation crowd out self-prediction? *Erkenntnis* 57: 91–122.

Ramsey, F. P. 1990 [1926]. Truth and probability. In his *Philosophical Papers*, ed. D. H. Mellor. Cambridge: Cambridge University Press: 52–94.

1990 [1928]. Universals of law and of fact. In his *Philosophical Papers*, ed. D. H. Mellor. Cambridge: Cambridge University Press: 140–4.

1990 [1929]. General propositions and causality. In his *Philosophical Papers*, ed. D. H. Mellor. Cambridge: Cambridge University Press: 145–63.

Reichenbach, H. 1971. *The Direction of Time*. Berkeley: University of California Press.

Reid, T. 2001 [1792]. Of power. *Philosophical Quarterly* 51: 1–12.

Resnik, M. 1987. *Choices: An Introduction to Decision Theory*. Minneapolis: University of Minnesota Press.

Riker, W. H. and P. C. Ordeshook. 1968. A theory of the calculus of voting. *American Political Science Review* 62: 25–43.

Ritchie, A. D. 1967. *George Berkeley: A Reappraisal*. Manchester: Manchester University Press.

Rosen, D. 1978. In defense of a probabilistic theory of causality. *Philosophy of Science* 45: 604–13.

Ross, L., D. Greene and P. House. 1977. The 'false consensus effect': an egocentric bias in social perception and attribution processes. *Journal of Experimental Social Psychology* 13: 279–301.

Russell, B. A. W. 1913. On the notion of cause. *Proceedings of the Aristotelian Society* 13: 1–26.

1926 [1914]. *Our Knowledge of the External World*. London: George Allen and Unwin.

Salmon, W. 1984. *Scientific Explanation and the Causal Structure of the World*. Princeton: Princeton University Press.

Samuelson, P. 1963. Risk and uncertainty: a fallacy of large numbers. *Scientia* 98: 108–13.

Sartre, J. 1969. *Being and Nothingness*. Tr. H. Barnes. London: Routledge.

Savage, L. J. 1972. *The Foundations of Statistics*. 2nd edition. New York: Dover.

Schopenhauer, A. 1995 [1819]. *The World as Will and Idea*. Abridged. Ed. D. Berman. Tr. J. Berman. London: Everyman.

Seidenfeld, T. 1984. Comments on Causal Decision Theory. *Philosophy of Science Association* 2: 201–12.

Sen, A. K. 1977. Rational fools: a critique of the behavioural foundations of economic theory. *Philosophy and Public Affairs* 6: 317–44.

Shafir, E. and A. Tversky. 1992. Thinking through uncertainty: nonconsequential reasoning and choice. *Cognitive Psychology* 24: 449–74.

Shafir, S., T. Reich, E. Tsur, I. Erev and A. Lotem. 2008. Perceptual accuracy and conflicting effects of certainty on risk-taking behaviour. *Nature* 453: 917–20.

Skyrms, B. 1980. *Causal Necessity*. New Haven: Yale Universty Press.

1982. Causal Decision Theory. *Journal of Philosophy* 79: 695–711.

1984a. EPR: Lessons for metaphysics. *Midwest Studies in Philosophy* 9: 245–55.

1984b. *Pragmatics and Empiricism*. New Haven: Yale Universty Press.

Slezak, P. 2013. Realizing Newcomb's problem. Available online from philsci-archive.pitt.edu (accessed 2 July 2013).

Sloman, S. 2005. *Causal Models*. Oxford: Oxford University Press.

Sobel, J. H. 1986. Notes on decision theory. *Australasian Journal of Philosophy* 64: 407–37. Reprinted in his *Taking Chances*. Cambridge: Cambridge University Press, 1994: 141–73.

1988. Infallible predictors. *Philosophical Review* 97: 3–24. Reprinted in his *Taking Chances*. Cambridge: Cambridge University Press, 1994: 100–18.

1989. Kent Bach on good arguments. *Canadian Journal of Philosophy* 20: 447–53.

1990. Newcomblike problems. *Midwest Studies in Philosophy* 15: 224–55. Reprinted in his *Taking Chances*. Cambridge: Cambridge University Press, 1994: 31–76.

Stalnaker, R. 1980 [1972]. Letter to David Lewis. In W. L. Harper, R. Stalnaker and G. Pearce (eds.), *Ifs: Conditionals, Belief, Decision, Chance, and Time*. Dordrecht: D. Reidel: 151–2.

Strotz, R. H. 1955. Myopia and inconsistency in dynamic utility maximization. *Review of Economic Studies* 23: 165–80.

Suppes, P. 1970. *A Probabilistic Theory of Causality*. Amsterdam: North-Holland.

Valentini, A. 2009. Beyond the quantum. *Physics World* 2009: 32–7.

Van Fraassen, B. C. 1991. *Quantum Mechanics: An Empiricist View*. Oxford: Clarendon Press.

Van Inwagen, P. 1983. *An Essay on Free Will*. Oxford: Oxford University Press.

Varian, H. 1992. *Microeconomic Analysis*. 3rd edition. New York: W. W. Norton and Co.

2003. *Intermediate Microeconomics*. 6th edtion. New York: W. W. Norton and Co.

Verbeek, B. 2008. Consequentialism and rational choice: lessons from the Allais paradox. *Pacific Philosophical Quarterly* 89: 86–116.

Von Wright, G. H. 1973. On the logic and epistemology of the causal relation. In P. Suppes et al. (eds.), *Logic, Methodology and Philosophy of Science IV*. Amsterdam: North-Holland: 293–312. Reprinted in E. Sosa and M. Tooley (eds.), *Causation*. Oxford: Oxford University Press, 1993: 105–24.

Wallace, D. 2007. Quantum probability from subjective likelihood: improving on Deutsch's proof of the probability rule. *Studies in History and Philosophy of Modern Physics* 34: 415–39.

Weber, M. 1992 [1920]. *The Protestant Ethic and the Spirit of Capitalism*. Tr. T. Parsons. London: Routledge.

Wegner, D. 2002. *The Illusion of Conscious Will*. Bradford Books, Cambridge, MA: MIT Press.

Weirich, P. 1998. *Equilibrium and Rationality: Game Theory Revised by Decision Rules*. Cambridge: Cambridge University Press.

2001. *Decision Space: Multidimensional Utility Analysis*. Cambridge: Cambridge University Press.

2013. Decisions without sharp probabilities. Unpublished typescript.

Williamson, J. 2008. Objective Bayesian probabilistic logic. *Journal of Algorithms* 63: 167–83.

Williamson, T. 2007. *The Philosophy of Philosophy*. Oxford: Blackwell.

Wittgenstein, L. 2009 [1951]. *Philosophical Investigations*. Tr. G. E. M. Anscombe, P. M. S. Hacker and J. Schulte. Revised 4th edition by P. M. S. Hacker and J. Schulte. Malden: Wiley-Blackwell.

Wolfson, S. 2000. Students' estimates for the prevalence of drug use: evidence for a false consensus effect. *Psychology of Addictive Behaviours* 14: 295–8.

Wolpert, D. H. and G. Benford. 2013. The lesson of Newcomb's paradox. *Synthese* 190: 1637–46.

Woodward, J. 2003. *Making Things Happen*. Oxford: Oxford University Press.

Index

Printed in the United States
By Bookmasters